LUMINESCENT SPECTROSCOPY of PROTEINS

LUMINESCENT SPECTROSCOPY of PROTEINS

Eugene A. Permyakov
Senior Researcher
Institute of Biological Physics of Russia
Russian Academy of Sciences
Puchino, Moscow
Russia

CRC Press
Boca Raton Ann Arbor London Tokyo

Library of Congress Cataloging-in-Publication Data

Permiakov, E. A. (Evgenii Anatol'evich)
Luminescent spectroscopy of proteins / author, Eugene A. Permyakov.
p. cm.
Includes bibliographical references and index.
ISBN 0-8493-4553-7
1. Proteins—Analysis. 2. Proteins—Spectra. 3. Luminescence spectroscopy. I. Title.
QP551.P397 1992
574.19'245—dc20 92-19189
CIP

Direct all inquiries to CRC Press, Inc., 2000 Corporate Blvd., N.W., Boca Raton, Florida, 33431.

International Standard Book Number 0-8493-4553-7

Library of Congress Card Number 92-19189

Printed in the United States 1 2 3 4 5 6 7 8 9 0

Printed on acid-free paper

The Author

Eugene Anatolyevich Permyakov was born in Ivanovo (U.S.S.R.) in 1946. In 1970 he graduated from the Moscow Physical Technical Institute, specializing in molecular biophysics. Since 1970 he has worked at the Institute of Biological Physics of the U.S.S.R. Academy of Sciences in Pushchino (Moscow region) as a senior researcher.

In 1976 he defended his candidate dissertation (similar to a Ph.D. in Western countries) which was devoted to luminescent study of freezing of molecular dynamics of proteins. In 1988 he defended his doctoral dissertation (to get the top scientific degree in the U.S.S.R.). The dissertation dealt with physico-chemical and functional properties of calcium-binding proteins which were the main objects of his studies. Dr. Permyakov's main experimental method is the method of intrinsic protein luminescence. He has published one monograph (*Parvalbumin and Related Calcium Binding Proteins,* Moscow, Nauka, in Russian) and more than 60 articles.

Dr. Permyakov was a visiting professor at the Ohio State University in Columbus, Ohio in September 1990 for one year.

Table of Contents

Chapter 4

Chapter 1

Introduction

Luminescence is defined as an emission of photons from electron-excited states of molecules. There are two types of luminescence: fluorescence and phosphorescence. Fluorescence arises due to radiative transitions between singlet states of a molecule, while phosphorescence is caused by radiative transitions between triplet and singlet states.

The works of many scientists, including Debye and Edwards,[1] Weber,[2] Szent-Györgyi,[3] Duggan and Udenfriend,[4] Shore and Pardee,[5] Vladimirov,[6] Konev,[7] Burstein,[8,9] and many others, have shown that proteins in aqueous solutions possess intrinsic luminescence in the near-ultraviolet (UV) region which can be used for structural and physicochemical studies of proteins.

Aromatic amino acid residues of tryptophan, tyrosine, and phenylalanine are the main luminescent groups in proteins. Luminescence parameters reflect properties of excited states, characteristics of electronic transitions in a luminophore, and interactions of the luminophore with its environment. Studies of protein luminescence provide basic information for understanding photochemical reactions in proteins. Luminescence parameters allow one to determine localization of the aromatic amino acid residues in proteins and to characterize their environment. Using very high sensitivity of luminescence to changes in environment of luminophores we can study structural and physicochemical changes in protein molecules, including functionally significant changes. The inertness of this method permits kinetic registration of the changes.

The intrinsic luminescence method is now one of the most popular and widespread methods in biophysics and biochemistry. Due to the relative cheapness of spectrofluorimeters (especially for steady-state measurements), this equipment is rather common for many biophysical and biochemical laboratories.

This book is intended for students and scientists in biophysics, biochemistry, and biology who would like to apply the intrinsic luminescence method for their experimental studies of proteins. It represents an attempt to fill the gap in modern spectroscopic literature between very general information about protein luminescence in most textbooks and highly specialized books and reviews for experts in this field.

The main aim of this book is to demonstrate the basic principles of the intrinsic fluorescence method and to teach how to use it in biophysical and biochemical studies. The book is devoted mostly to steady-state luminescence, though it contains some information about time-resolved luminescence spectroscopy. The reason for this is the fact that the sophisticated instrumentation for time-resolved or frequency-domain fluorescence measurements is usually very expensive and inaccessible to most biophysical and biochemical laboratories which are not specialized in luminescence spectroscopy. The book does not consider the fluorescence polarization method, which deserves a separate book. It contains some general information on the main principles of instrumentation for luminescence spectroscopy, but it does not consider it in detail, since modern industry produces many types of various spectral equipment; their consideration would require a separate book as well.

The main part of the experimental material for this book was taken from the rich experience of the Group of Spectroscopy of the Institute of Theoretical and Experimental Biophysics of the Russian Academy of Sciences, headed by Professor Edward A. Burstein. Most of these data were published in various original articles, but they have never been collected together in a book or review. The book does not duplicate the information contained in well-known textbooks on protein luminescence (see for example the excellent book by Lakowicz[10]), but provides additional material.

The book consists of three main parts. The first part deals with general physical principles of the luminescence method and requires a minimal knowledge of quantum theory of atoms and molecules. It describes the basic photophysical processes in organic molecules: fluorescence, phosphorescence, internal conversion, vibrational relaxation, intersystem crossing, and so forth, and gives information about effects of intermolecular interactions on luminescence parameters, such as fluorescence spectrum position and fluorescence quantum yield. This part is supposed to give a physical base for the other parts of the book and to make the book autonomous. Since the intrinsic fluorescence method belongs to physical methods, which are impossible without mathematics, I was forced to use some simple mathematical equations which, I believe, will be helpful in understanding the material better.

The second part discusses spectral properties of the main protein chromophores, tryptophan, tyrosine, and phenylalanine in the isolated state: their absorption and emission spectra at various conditions, fluorescence quenching by protein groups, emission lifetimes, and so on. This information is needed for interpretation of spectral data for proteins.

The third part describes protein luminescence and its use for studies of structural, physicochemical, and functional properties of proteins. This part demonstrates what kind of information can be obtained from an analysis of position and shape of protein fluorescence spectra, fluorescence quantum yields of proteins, and fluorescence decay of proteins. It discusses general tactics of luminescent studies of proteins and methods of interpretation of spectral data. It shows what fluorescence parameters should be used for obtaining quantitative information. This chapter deals with studies of dependencies of protein fluorescence parameters on temperature, pH, ionic strength, and denaturant concentration. It also discusses how to study interactions of proteins with low-molecular-mass compounds by the intrinsic fluorescence method.

It should be noted once again that this book does not compete with the well-known books and reviews on protein fluorescence. In a sense, it is complementary to them, and if a reader wants to study the luminescence method in more detail, it is suggested that this book be used in combination with other sources, especially the well-known book by Lakowicz.[10]

REFERENCES

1. Debye, P. and Edwards, J. O., *Science,* 116, 143, 1952.
2. Weber, G., *Adv. Protein Chem.,* 8, 415, 1953.
3. Szent-Györgyi, A., *Biochim. Biophys. Acta,* 16, 195, 1955.
4. Duggan, D. E. and Udenfriend, S., *J. Biol. Chem.,* 223, 313, 1956.
5. Shore, V. G. and Pardee, A. B., *Arch. Biochem. Biophys.,* 60, 100, 1956.
6. Vladimirov, Y. A., *Photochemistry and Luminescence of Proteins,* Nauka, Moscow, 1965.
7. Konev, S. V., *Fluorescence and Phosphorescence of Proteins and Nucleic Acids,* Plenum Press, New York, 1965.
8. Burstein, E. A., *Luminescence of Protein Chromophores,* Vol. 6, Model Studies, Science and Technology Results. Biophysics, VINITI, Moscow, 1976.
9. Burstein, E. A., *Intrinsic Protein Luminescence,* Vol. 6, Origin and Applications. Science and Technology Results. Biophysics, VINITI, Moscow, 1977.
10. Lakowicz, J. R., *Principles of Fluorescence Spectroscopy,* Plenum Press, New York, 1990.

Chapter 2

Energy Levels in Molecules and Transitions Between Them

I. INTRODUCTION

It is well known from quantum mechanics[1-3] that in order to describe a molecule, one should solve the corresponding Schrödinger equation, i.e., to find the eigenvalues E_i and eigenfunctions Ψ_i of the Hamiltonian operator:

$$H\Psi = E\Psi \tag{1}$$

where H is a sum of the operators for kinetic energy of electron motion T_e, kinetic energy of nuclear motion T_n, and potential energy of the interactions between electrons, between nuclei, and between electrons and nuclei V:

$$H = T_e + T_n + V \tag{2}$$

The Hamiltonian depends on both nuclear coordinates X and electron coordinates x:

$$H(x, X)\Psi(x, X) = E\Psi(x, X) \tag{3}$$

Equation 3 does not take into account the spins of electrons and is valid for particles wherein the velocities of which are low in comparison with the velocity of light. Motions of electrons are usually assumed to be dependent only on nuclear coordinates and not on nuclear velocities. This is usually the case since the masses of electrons are much less than the masses of nuclei. Such an approach is called the *adiabatic* or *Born-Oppenheimer approximation.* In this approximation the eigenfunction $\Psi(x,X)$ of a molecule can be represented as a product of the electron $\Psi_e(x,X)$ and nuclear $\Psi_n(X)$ eigenfunctions:

$$\Psi(x,X) = \Psi_e(x,X)\Psi_n(X) \tag{4}$$

The electron eigenfunction $\Psi_e(x,X)$ is determined from the electron Schrödinger equation describing motions of electrons for a fixed configuration of nuclei:

$$H_e(x,X)\Psi(x,X) = E_e(X)\Psi_e(x,X) \tag{5}$$

The Hamiltonian is given by

$$H_e = T_e + V_e = [-h^2/8\pi^2 m]\Sigma\partial^2/\partial x_i^2 + V_e(x,X) \tag{6}$$

where V_e is the operator of potential energy for interactions of electrons with each other and with nuclei, m is mass of electron, and h is Plank's constant.

The nuclear eigenfunction Ψ_n, describing nuclear motions, is a solution of the nuclear Schrödinger equation:

$$H_n(X)\Psi_n(X) = (E - E_n)\Psi_n(X) \tag{7}$$

where H_n is given by

$$H_n = T_n + V_n = [-h^2/8\pi^2]\Sigma[1/M_k]\partial^2/\partial X_k^2 + V_n \tag{8}$$

M_k and X_k are the mass and coordinates of the kth nucleus. The operators V_n, V_e, and V are related by the following equation:

$$V = V_e + V_n \tag{9}$$

Usually the total energy is represented as a sum of the electron energy, computed at a fixed nuclear configuration, $E_e(X_0)$, and the nuclear energy E_n:

$$E = E_e(X_0) + E_n \tag{10}$$

Spectroscopy provides confirmation of the validity of the adiabatic approximation. In the infrared (IR) spectra, we measure transitions between different vibrational levels of the same electronic state. This pattern of the vibrational spectrum can be described by the eigenfunctions $\Psi_n(X)$ and eigenenergies E_n. In the visible and UV regions of the spectrum, we measure transitions between different electronic levels, and the resulting spectrum contains structures arising from different vibrational components of the electronic states.

The eigenfunction of a system of particles possessing spins should depend on both spatial coordinates and spin variables. Upon weak spin-orbital coupling the total electronic wave function can be represented as a product:

$$\Psi_e = \Psi_c \cdot \Psi_s \tag{11}$$

where Ψ_c is the coordinate wave function and Ψ_s is the spin wave function.

It is well known that the electron spin s = $^1/_2$. Particles with such spin obey the so-called Pauli exclusion principle, i.e., the total electron wave function must be antisymmetrical with respect to rearrangement of any electron pair.

The electron wave function Ψ_e describes the electron cloud in a molecule: electron density at any point in space is proportional to $|\Psi_e|^2$. For each electron one can show a boundary surface which contains the greater part of the electron cloud. This surface is a clear image of the electron orbital. Complex molecules possess a large number of such orbitals, which are described by wave functions Ψ_e. The Pauli's exclusion principle means that two, and only two, electrons may occupy a molecular orbital, and these electrons must have opposite spin functions.

So in order to compute electronic spectrum of a molecule, one should solve the electronic Schrödinger equation (Equation 5), i.e., find electronic eigenfunctions $\Psi_e(x,X)$ and eigenvalues $E_e(X)$ of the energy operator. The set of the eigenvalues of the energy operator correspond to a succession of possible energy levels in the molecule. Apart from this, one should also know selection rules and intensities of transitions.

The knowledge of a wave function Ψ allows one to obtain the eigenvalue E:

$$E = \frac{\int \Psi^* H \Psi \, d\tau}{\int \Psi^* \Psi \, d\tau} \tag{12}$$

where Ψ^* is the complex conjugate of Ψ.

The computation of wave function and eigenvalues of electronic energy for multiatomic molecules is very complicated. The most successful results are obtained for the simplest diatomic molecules.

II. ENERGY LEVELS

An electron transition induced by absorption of a photon by a molecule can be represented by a scheme of energy levels.[2-6] Such schemes usually show energies of the lowest vibrational levels for each excited state with respect to the lowest vibrational level of the ground state. Sometimes higher vibrational levels are also shown. Figure 1 demonstrates a general scheme of the lowest energy levels of an organic molecule. On the left part of the Figure is shown a set of singlet states, S_k, i.e., the states with zero total spin. On the right part of the Figure is shown the scheme of triplet states, T_k, i.e., the states for which the total spin of the molecule equals 1.

Triplet state T_k possesses the same electronic configuration as a corresponding singlet state S_k. These states differ by the relative orientation of the spins of two electrons on the outer orbital. According to Hund's rule, the triplet energy level is always lower than the corresponding singlet one. The lowest lines for each S_k or T_k state in Figure 1 denote purely electronic levels. Actually

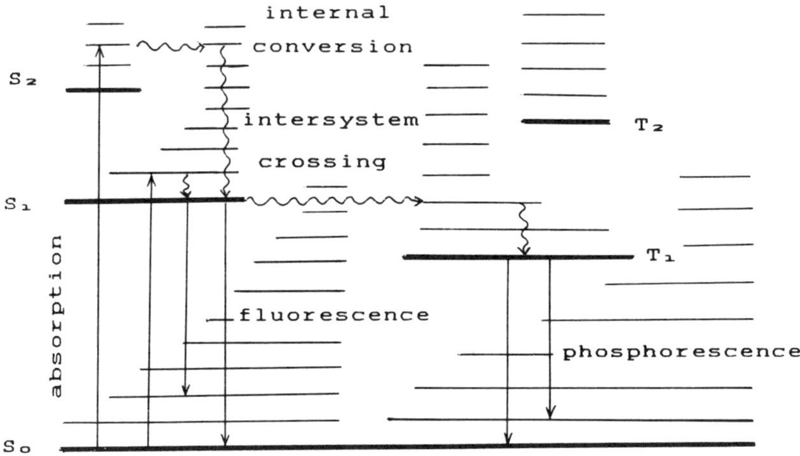

Figure 1. General scheme of the lowest energy levels of an organic molecule.

these lines correspond to hypersurfaces of potential energy in a multidimensional space. Each hypersurface can intersect with any other hypersurface. Energy in a set of states increases with an increase in the state number. S_0, S_1, S_2, . . . ,S_i, . . . are arranged in the order of increasing of energy.

III. INTERNAL CONVERSION AND VIBRATIONAL RELAXATION

Absorption of a light quantum, depending on its energy, transfers a molecule from the ground state, S_0, to one of the vibrational levels of the excited states, S_1, S_2, and so on. It was found empirically (Kasha rule) that in condensed media the excitation energy of the S_2 state or any other higher state quickly dissipates due to inelastic collisions and the molecule nonradiatively reaches the zero vibrational level of the S_1 state. Nonradiative transitions between electronic states of the same spin multiplicity are called *internal conversion*. In the internal conversion process the energy is transferred from a point on the potential energy hypersurface of a state (S_2, for example) to the potential energy hypersurface of S_1. After that, the excess vibrational energy which is possessed by the quasidegenerated S_1 state dissipates in the course of vibrational relaxation and the system reaches the thermally equilibrated state of S_1 molecules. In this process the excitation energy is transferred to the external medium. Nonradiative deactivation $S_1 \rightsquigarrow S_0$ follows this mechanism too. It is assumed that the rate constants of internal conversion and vibrational relaxation exceed 10^{12} s^{-1} (Figure 2).

IV. FLUORESCENCE

Radiative transitions in a molecule between levels of the same multiplicity give rise to *fluorescence*. It is evident that the emission occurs as a result of the transition of the molecule from the S_1 to the S_0 state. A comparison of the processes of fluorescence $S_0 \leftarrow S_1$ and absorption $S_0 \rightarrow S_1$ (Figure 3) shows that the longest wavelength line of absorption coincides in energy with the shortest wavelength line of emission. This line is called the 0-0 line. Let's formulate the main properties of molecular fluorescence.

1. Due to internal conversion, and in accordance with the Kasha rule, the position of the fluorescence spectrum does not depend on the excitation wavelength.
2. Fluorescence spectrum is shifted towards longer wavelengths with respect to the absorption band $S_0 \rightarrow S_1$ (Stocks rule) and is approximately a mirror image of this band.
3. The number of quanta emitted per unit time is proportional to the number of quanta absorbed per unit time and fluorescence quantum yield:

$$N_f = I_0(1-10^{-D})\cdot q_f \quad (13)$$

where I_0 is intensity of incident light; D is the absorbance of the solution, which is the product of the extinction coeffi-

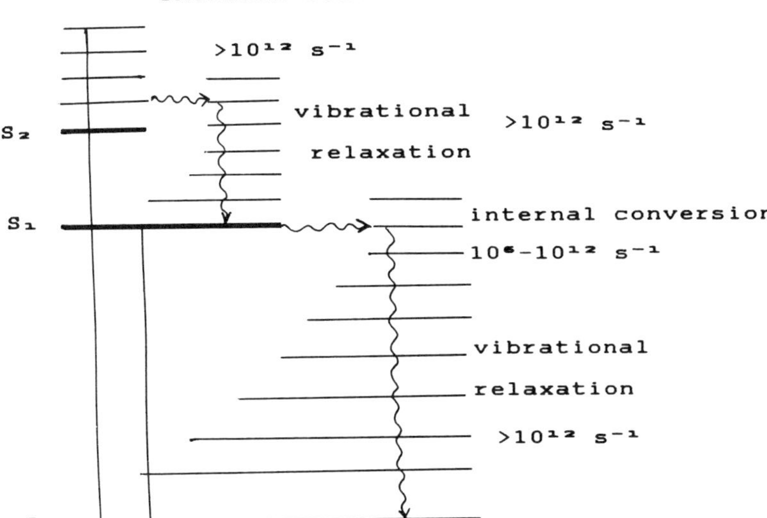

Figure 2. Processes of internal conversion and vibrational relaxation.

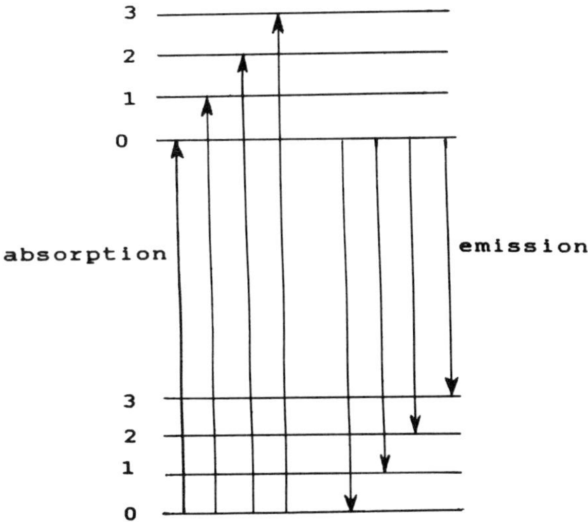

Figure 3. Electron-vibrational transitions $S \leftrightarrow S_0$.

cient of the molecule ε, the concentration of the molecules c, and the optical path length l ($D = \varepsilon \cdot c \cdot l$); q_f is *fluorescence quantum yield,* which is the ratio of the number of quanta emitted from an excited state to the number of quanta absorbed during the transitions from the ground to this excited state per time unit. If concentration c is low, then

$$N_f = 2.303 \cdot \varepsilon \cdot c \cdot l \cdot I_0 \cdot q_f \tag{14}$$

4. If the fluorescence quantum yield does not depend on excitation wavelength λ_{ex}, then for dilute solutions fluorescence intensity at excitation at λ_{ex} $I_f(\lambda_{ex})$ is proportional to $\varepsilon(\lambda_{ex}) \cdot I_0(\lambda_{ex})$. If I_0 also does not depend on λ_{ex}, then $I_f(\lambda_{ex})$ is proportional to $\varepsilon(\lambda_{ex})$ i.e., the spectra of excitation of fluorescence and of absorption are identical.

V. INTERSYSTEM CROSSING

Nonradiative transitions between electron states of different multiplicity are called *intersystem crossing*. For example, it is possible to have $S_1 \rightsquigarrow T_1$ and $T_1 \rightsquigarrow S_0$ intersystem crossing (Figure 4). The process of intersystem crossing requires change of spin, since it connects states of different spin multiplicity. In the course of internal conversion spin does not change. Any transition, radiative or nonradiative, between states of different multiplicity is forbidden. This forbiddance is partially eliminated by an interaction between magnetic dipoles arising as a result of the spin and orbital movements of the electron. Such an interaction is called spin-orbit interaction. The energy of the spin-orbit interaction depends upon the relative orientation of the spin and orbit quantum mechanical moments of electrons, and the classical operator for the spin-orbit interaction of a single electron in central potential field is:

$$H_{s0} = k \cdot \beta \cdot (LS) \tag{15}$$

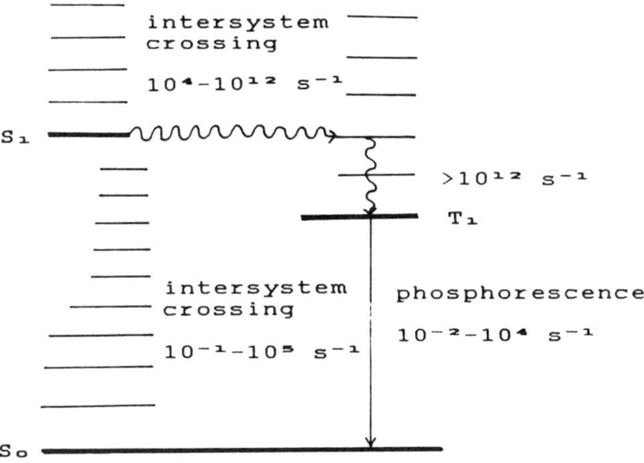

Figure 4. Processes of intersystem crossing $S_1 \rightsquigarrow T_1$ and $T_1 \rightsquigarrow S_0$.

where β is a parameter depending on the potential field of nuclei, L is the orbital angular momentum operator of the electron, S is the spin angular momentum operator of the electron, and k is a coefficient constant for a given molecule. Taking into consideration the spin-orbit coupling, one can write for the wave function

$$\Psi_{s0} = \Psi_t^0 + \lambda \Psi_s^0 \tag{16}$$

where Ψ_t^0 and Ψ_s^0 are wave functions of pure unperturbed triplet and singlet states. λ is a constant reflecting the contribution of the singlet state:

$$\lambda = \left| \frac{\int \Psi_s^0 H_{so} \Psi_t^0 \, d\tau}{E_t - E_s} \right| \tag{17}$$

where E_t is the energy of the triplet state Ψ_t^0, and E_s is the energy of singlet state Ψ_s^0. From Equations 15 to 17 one can isolate a part of the wave function Ψ_{s0} reflecting the singlet-triplet transition:

$$\Psi_{s0} = \frac{\beta \int \Psi_s^0 (LS) \Psi_t^0 \, d\tau}{|E_t - E_s|} \Psi_s^0 \tag{18}$$

It is clearly seen that the value of the spin-orbit coupling is inversely proportional to the difference in energies of triplet and perturbed singlet states.

VI. PHOSPHORESCENCE

Radiative transition between two states of different multiplicity is called *phosphorescence.* An example of such transition is the transition $S_0 \leftarrow T_1$ (Figure 4). The long lifetime of a phosphorescent state, which is within the range of 10^{-4} to 10^2 s, is due to the requirement of spin reorientation and the low probability of this process. Due to a long lifetime of the state T_1, phosphorescence is especially sensitive to quenching; therefore, it is usually found at low temperatures, for example, at the temperature of liquid nitrogen, –196°C. Spin-orbit interaction affects both intersystem crossing and phosphorescence processes. Upon increase of the spin-orbit interaction, one can expect an increase in population of the T_1 state, an increase in the phosphorescence quantum yield, and a decrease in phosphorescence lifetime.

VII. FRANK-CONDON PRINCIPLE

The Frank-Condon principle is formulated as follows:[2-4] electron transitions are so fast (10^{-15} s) in comparison to the movements of nuclei (10^{-12} s), that during electron transitions nuclei can change neither their velocities nor their positions. The principle reflects the fact of impossibility of a fast conversion of electronic energy into energy of nuclear vibrations. The consequences of the Frank-Condon principle are best seen by consideration of potential energy curves. For a description of a multiatomic molecule, one should know a set of multidimen-

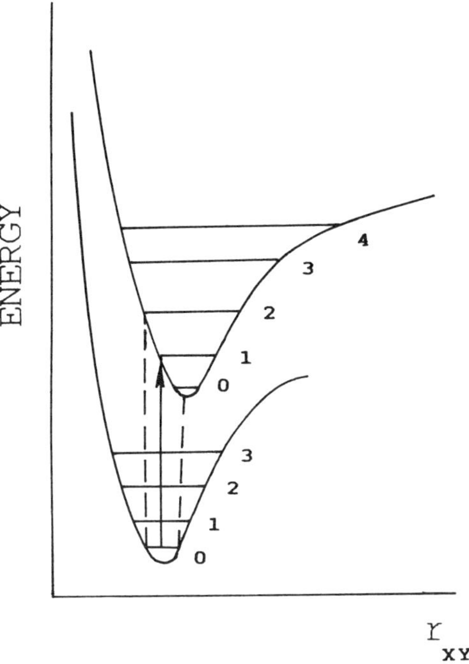

Figure 5. Potential energy curves for the ground and excited states of a diatomic molecule, XY.

sional energy surfaces. It is much simpler to consider the potential curve for a diatomic molecule. Conclusions drawn from its study can be used in generalizations for multiatomic molecules.

Figure 5 shows potential curves for the ground and excited states of a diatomic molecule. Quantum mechanical solution of the problem of harmonic and anharmonic oscillators provides the important result that not all energies are allowed for the vibrating molecule. For example, for the harmonic vibrations, only energies described by the equation

$$E = h\nu(v + 1/2) \tag{19}$$

(v is vibrational quantum number, which can be 0, 1, 2, . . .; ν is frequency of the harmonic oscillator vibrations; and h is Plank's constant) are allowed. It is clearly seen from this equation that the energy of the lowest vibrational state (v = 0) does not equal zero, but is hν/2. In real molecules the vibrations are not harmonic and their potential energy curves have the appearance shown in Figure 5. Horizontal lines on these curves show the allowed vibrational levels of the molecule.

Let's consider first of all transitions between electron vibrational levels of a molecule and the absorption of a light quantum. We assume that the molecule absorbing light is on the lowest vibrational level of the ground state. According to the Frank-Condon principle, distances between nuclei and their velocities should not change during the electron transition, i.e., the absorption is allowed only to vibrational levels in the region limited by the dotted lines. Vertical transitions intersect the upper potential curve at the turning point, where nuclei are stopped for a short period. Transitions to other vibrational levels are also possible if, during the absorption, the positions of nuclei or their velocities essentially change. So by means of the Frank-Condon principle we can predict probabilities of transitions between various levels. Most probable transitions are those that do not require changes of either internuclear distances or velocities of nuclei. Figure 6 shows intensities of the absorption transitions between various levels.

Radiative transition occurs from the zero vibrational level of the lowest excited state (Figure 7). The most probable transitions in this case are the vertical radiative transitions. Since the minimum of the potential energy curve for the excited state is shifted to the right from the minimum of the potential curve for the ground state, the molecule formed by virtue of the most probable radiative transitions will be extended, while the absorption results in a contracted molecule. Emission spectrum of a molecule with potential energy curve presented in Figure 7 is shown in Figure 8.

VIII. LUMINESCENCE LIFETIME

The lifetime, τ, of an excited state is the time required for the e-fold decrease of the initial number of excited states. The intrinsic lifetime, τ^0, of an excited state is a mean lifetime of this state

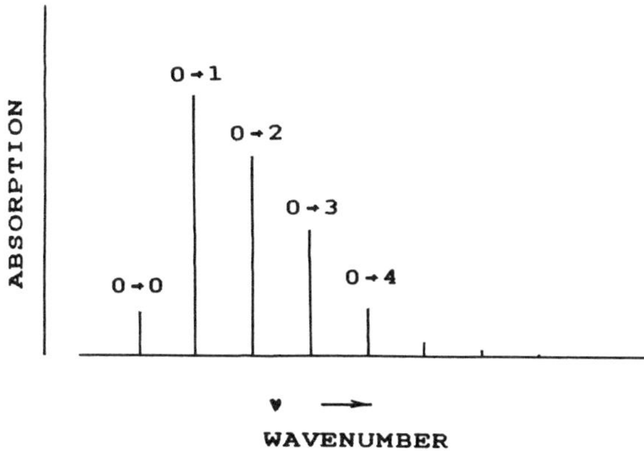

Figure 6. Absorption spectrum of the diatomic molecule for which the potential energy curves are presented in Figure 5.

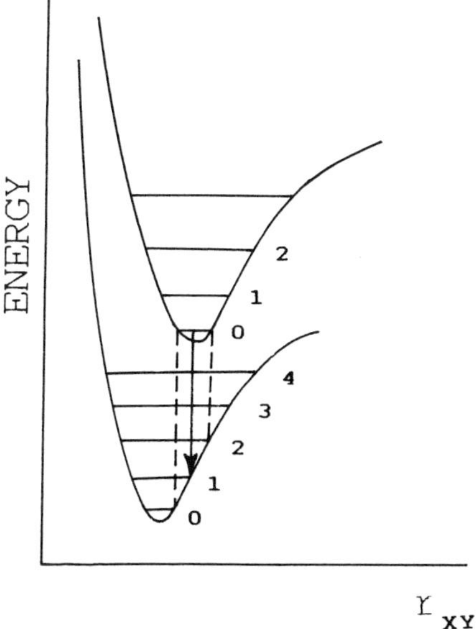

Figure 7. Potential energy curves for a diatomic molecule, XY.

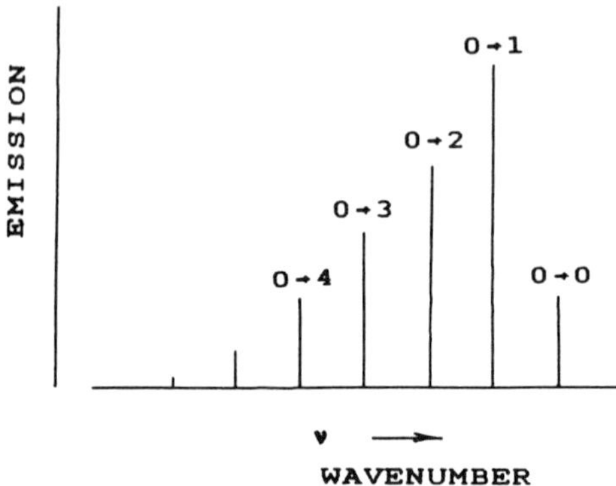

Figure 8. Emission spectrum of the diatomic molecule for which potential energy curves are shown in Figure 7.

in the absence of any nonradiative deactivation processes. If the nonradiative deactivation of the electron excitation occurs in the T_1 state and not in the S_1 state, then

$$q_f \cdot \tau_f^0 = \tau_f \tag{20}$$

$$\tau_p \frac{1 - q_f}{q_p} = \tau_p^0 \tag{21}$$

where q is quantum yield, τ^0 is intrinsic lifetime, and τ is the measured lifetime. Indices f and p refer to fluorescence and phosphorescence, respectively.

Let's consider the effects of competition of nonradiative processes on lifetime τ. Let A_0 and A^* be the ground and excited states of a molecule:

$$A^* \xrightarrow{k_1} A_0 + h\nu \tag{22}$$

$$A^* \overset{k_2}{\rightsquigarrow} A_0 \tag{23}$$

k_2 includes internal conversion, intersystem crossing, chemical reactions, and so on. The rate of disappearance of A* is

$$-\frac{d[A^*]}{dt} = (k_1 + k_2)[A^*] \tag{24}$$

$$[A^*] = [A^*]_0 \exp[-(k_1 + k_2)t] \tag{25}$$

i.e.,

$$\tau = 1/(k_1 + k_2) \tag{26}$$

or

$$\tau = 1/\Sigma k_i \tag{27}$$

where k_i are rate constants of monomolecular or pseudomonomolecular processes of deactivation of the A* state.

IX. FLUORESCENCE QUANTUM YIELD

Figure 9 shows a simplified scheme of electronic levels in a molecule. Straight arrows show radiative transitions, while wavy arrows show nonradiative transitions. k_f and k_p are fluorescence and phosphorescence rate constants, respectively; k_1 is the rate constant of the $S_1 \rightsquigarrow S_0$ internal conversion; k_{st} and k_{ts} are rate constants of $S_1 \rightsquigarrow T_1$ and $T_1 \rightsquigarrow S_0$ intersystem crossing, respectively. Lifetimes of the lowest excited singlet and triplet states are long enough to allow chemical reaction or nonradiative deactivation of the excited state (luminescence quenching) upon collision with other molecules. We shall consider only the quenching in singlet states. Fluorescence quenching is a second-order process, but since concentration of a quencher is usually much higher than the concentration of the excited molecules, it can be characterized by a first-order rate constant $k_q[Q]$, where k_q is the bimo-

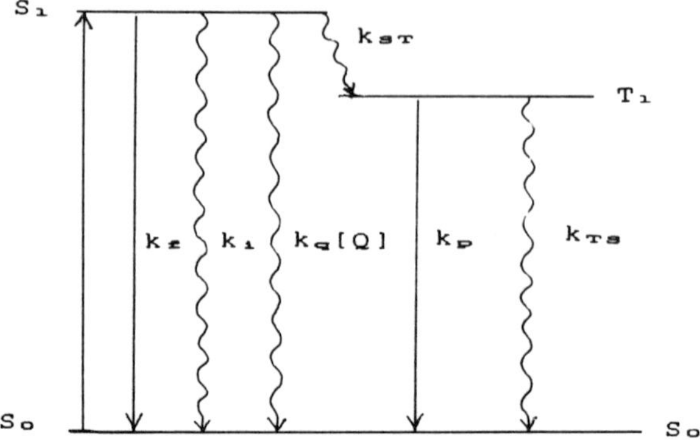

Figure 9. Simplified scheme of energy levels of a molecule.

lecular rate constant and [Q] is quencher concentration. Upon continual illumination of the solution of a period of time which is long enough in comparison with the fluorescence lifetime, the system will reach a steady state when the rate of creation of the singlet excited molecules (S_1) equals the rate of their disappearance in various processes.

In the absence of photochemical reactions in upper singlet states, the rate of S_1 formation is equal to the light absorption rate I_a. In the steady state

$$I_a = (k_f + k_i + k_{st} + k_q[Q]) \cdot [S_1] \qquad (28)$$

The total fluorescence emission rate is $k_f \cdot [S_1]$. According to definition, fluorescence quantum yield is the ratio of the number of quanta emitted per unit time to the total number of quanta absorbed during the same time period:

$$q = \frac{k_f \cdot [S_1]}{(k_f + k_i + k_{st} + k_q[Q]) \cdot [S_1]} \qquad (29)$$

i.e.,

$$q = \frac{k_f}{k_f + k_i + k_{st} + k_q \cdot [Q]} \qquad (30)$$

Taking into account Equation 27, we can write

$$q = k_f \cdot \tau \tag{31}$$

where

$$\tau = \frac{1}{k_f + k_i + k_{st} + k_q \cdot [Q]} \tag{32}$$

X. **STERN-VOLMER EQUATION**

In the absence of external quenchers, fluorescence quantum yield is

$$q_0 \frac{k_f}{k_f + k_i + k_{st}} \tag{33}$$

Then,

$$q_0 / q = 1 + \frac{k_q \cdot [Q]}{k_f + k_i + k_{st}} \tag{34}$$

$$q_0 / q = 1 + K_{sv} \cdot [Q] \tag{35}$$

where

$$K_{sv} = \frac{k_q}{k_f + k_i + k_{st}} = k_q \cdot \tau_0 \tag{36}$$

$$\tau_0 = \frac{1}{k_f + k_i + k_{st}} \tag{37}$$

τ_0 is fluorescence lifetime in the absence of a quencher.

Equation 35 is called the Stern-Volmer equation, and the constant K_{sv} is called the Stern-Volmer constant. Most often the Stern-Volmer equation is written in the following form:

$$q_0 / q - 1 = K_{sv} \cdot [Q] \tag{38}$$

The quenching capacity of a compound is characterized by the Stern-Volmer constant. K_{sv} can be determined from the slope of the $(q_0/q - 1)$ vs. [Q] plot.

It should be noted that the possibility of approximation of experimental data by the Stern-Volmer equation cannot serve as proof that the fluorescence quenching obeys the collision mechanism. We can obtain an equation similar to the Stern-Volmer one in the case of a quencher forming nonfluorescent complexes with fluorescent molecules. In this case, the constant of complex formation is used instead of the Stern-Volmer constant.

In the case of the collisional mechanism, the quenching event becomes possible when molecules of the fluorescent substance and the quencher meet each other. When they meet each other, the solvent molecules form a "cage" and prevent their separation. Therefore, before their separation they will undergo many collisions with each other and a result of one of these will be the fluorescence quenching. If the activation energy ΔE of this process is high, the rate constant of the reaction will be equal to the product of the collision number, steric coefficient, and exponential Boltzmann factor $\exp(-\Delta E/RT)$.

If the activation energy is low, the reaction will occur upon every meeting and its rate will depend only on diffusion rate. The bimolecular rate constant, k_c, of the diffusion-controlled process depends only on temperature and solvent viscosity:

$$k_c = A \cdot RT / \eta \tag{39}$$

where η is solvent viscosity, R is gas constant, T is absolute temperature, and A is a numerical coefficient. This equation was obtained from the more general Smoluchovski equation:

$$k_c = 4\pi\sigma_{AB}(D_A + D_B) \times 10^3 \tag{40}$$

and Stockes-Einstein equation:

$$D_A = RT / 6\pi r_A \eta \quad (41)$$

when

$$\sigma_{AB} = r_A + r_B = 2r_A \quad (42)$$

where σ_{AB} is collision distance, r_A and r_B are radii of A and B molecules, and D_A and D_B are diffusion coefficients of the reagents.

XI. EXCITATION ENERGY TRANSFER BETWEEN SINGLET STATES

There exist two types of long distance excitation energy transfer processes between singlet states of molecules.[4-6] The first one is emission of a light quantum by a donor molecule and its absorption by an acceptor molecule. Such a process takes place if the absorption spectrum of the acceptor overlaps the emission spectrum of the donor and the medium is transparent in this spectral region. This trivial effect often leads to artifacts in the obtainment of the correct fluorescence spectra.

The second process is a nonradiative transfer of excitation energy from donor to acceptor due to direct electrodynamic interaction between excited and unexcited molecules. The quantitative theory of excitation energy transfer of this type was developed by Förster. Conditions for the effective nonradiative energy transfer are good overlap of absorption spectrum of donor with fluorescence spectrum of acceptor and high fluorescence quantum yield of donor. Efficiency of the intermolecular dipole-dipole energy transfer is characterized by a critical distance R_c. R_c is a distance between the donor and acceptor molecules at which the probability of the transition equals the probability of spontaneous deactivation. According to Förster's theory,[7]

$$R_c^6 = \frac{900(\ln 10)\kappa^2 q_f}{128\pi^6 n^4 N} \int_0^\infty \frac{f_D \varepsilon_A d\nu}{\nu^4} \quad (43)$$

where N is Avogadro's number, n is the refractive index of the medium, κ^2 is orientation factor (for freely rotating molecules in solution it equals $^2/_3$), q_f is fluorescence quantum yield, f_D is fluorescence spectrum of donor normalized to unity, ε_A is absorption spectrum of acceptor, and ν is wavenumber.

XII. EFFECTS OF INTERMOLECULAR INTERACTIONS ON ELECTRON-VIBRATIONAL TRANSITIONS IN MOLECULES

Interactions of the absorbing and emitting molecules with solvent molecules in liquid solution essentially affect their electron-vibrational transitions.[8,9] Studies of the electron-vibrational fluorescence and absorption spectra allow one to obtain valuable information about the chromophore environment. According to the most widespread classification, all types of intermolecular interactions are divided into *universal* and *specific* ones. Universal interactions between molecules occur in all cases without exception and depend upon their physicochemical parameters. After configuration and space averaging, they characterize the collective effects of the environment on the properties of a given molecule. Specific interactions do not occur in all systems. They are individual, quasichemical, exchange interactions. These interactions are characterized by high selectivity to molecular properties and result in formation of rather strong links between molecules.

Description of the universal interactions requires consideration of the interactions of a given molecule with the surrounding molecular ensemble, i.e., the molecule in the condensed medium is characterized by a potential of collective interaction with the surrounding molecules. Analytical expressions for potentials exist only for pair interactions, i.e., for interactions of a given molecule with one of the surrounding molecules. The transition from a potential of pair interactions $\phi(r)$ to potential of collective interactions $\Phi(r)$ is achieved only by using liquid models and is connected with many difficulties.

The potential of the universal Van der Waals interaction of two molecules ϕ_{univ} is

$$\phi_{univ} = \phi_{or} + \phi_{ind} + \phi_{disp} \tag{44}$$

where ϕ_{or}, ϕ_{ind}, and ϕ_{disp} are potentials of orientation, induction, and dispersion interactions, respectively. In the condensed medium, each molecule is affected not by a single neighboring molecule, but by an ensemble of molecules. The simplest assumption of the potential of collective interactions Φ is the assumption as to the additivity of the total energy of interactions:

$$\Phi = \Sigma\phi_i(R) \tag{45}$$

the main contribution being given by the molecules of the first coordination sphere. Using the radial distribution function g(R), one can write the expression for Φ:

$$\Phi = C\int \phi(R)g(R)R^2\, dR \tag{46}$$

Unfortunately, application of Equation 46 to real systems is as yet practically impossible because of a poor knowledge of ϕ and g. This forces us to avoid these difficulties by creating simpler models which allow us to evaluate the value of Φ. As a rule, these models are based on the substitution of the real interactions in a system of particles by a self-consistent external field affecting each molecule. Many models which are used now arose as a result of development and modification of models of liquid state.

Let's consider the mechanisms of influence of the universal interactions in the liquid phase on the parameters of the electronic bands of organic chromophores. Let's assume that the emitting molecule (chromophore) possesses a dipole moment both in the ground and in the excited states. We shall consider a transition between two electron states with energies E_g (ground state) and E_e (excited state). Since the lifetime of the molecule in the ground state is infinitely long, solvent molecules around the

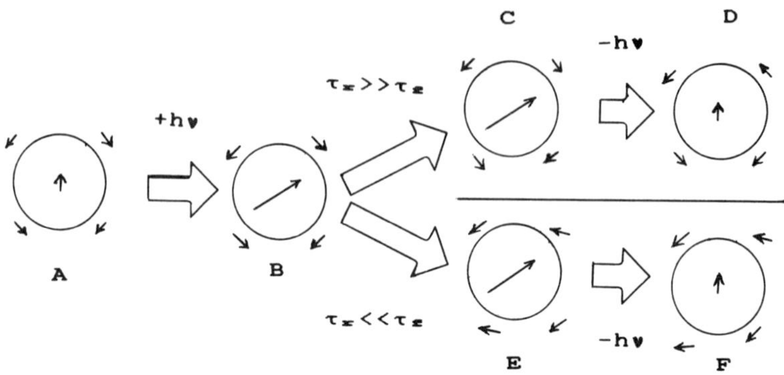

Figure 10. Processes of orientational relaxation of dipoles in the environment of a chromophore upon absorption and emission of light by a chromophore possessing a dipole moment (for explanations, see the text).

unexcited chromophore are oriented in such a way that the energy of the chromophore-solvent system is minimal (Figure 10A).

Absorption of a light quantum causes a transition of the molecule to the excited state (Figure 11). Due to a redistribution of electron density, this results in a change of the value and orientation of the dipole moment of the molecule (Figure 10B). According to the Frank-Codon principle, the electron transition occurs very rapidly so that the configuration of the solvent molecules around the chromophore does not significantly change during the electron transition. The energy level occupied by the molecule in this state is a nonequilibrium one; it is called the Frank-Condon level. After the electron transition, the solvent molecules start to reorient around the excited chromophore in such a way that their new configuration would be the most energetically favorable for the new electron density distribution. If the orientational relaxation time τ_r is much longer than the lifetime of the excited state of the chromophore τ_f, the relaxation has no time to occur and the emission will proceed from the practically nonequilibrium Frank-Condon state (Figure 10 C and D; Figure 11). The emission spectrum will have a shorter wavelength position in this case.

In the opposite case, when $\tau_r \ll \tau_f$, there is enough time for relaxation to be completed, and during the excitation lifetime, the

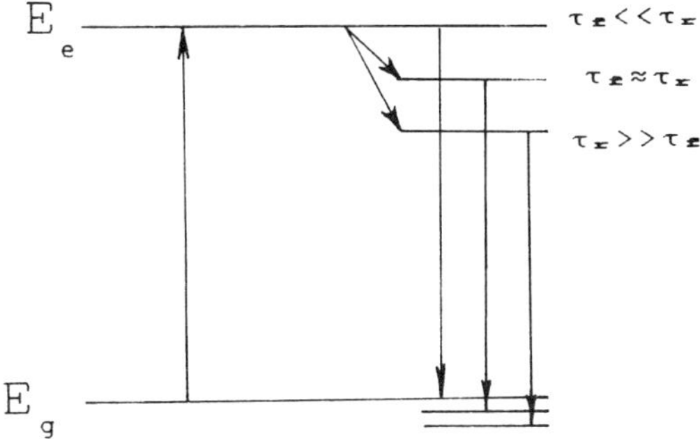

Figure 11. Scheme of electron levels in a molecule in a polar solvent for different relationships between fluorescence lifetime, τ_f, and relaxation time of the solvent molecules, τ_r.

solvent molecules reorient around the chromophore to the most favorable configuration. The emission will occur from the equilibrium-excited level corresponding to the most favorable orientation of the surrounding molecules around the excited molecule (Figure 10 F and G; Figure 11). The emission spectrum will have a longer wavelength position in this case.

In the case when τ_f is comparable with τ_r, the relaxation will be incomplete and the emission will occur from some intermediate level located between the levels corresponding to the situation when $\tau_f >> \tau_r$ and $\tau_f << \tau_r$. The position of the emission spectrum will be intermediate. Within the framework of the Debye model of relaxation, the position of the fluorescence spectrum at time t, ξ, is connected with the position of the spectrum at time zero and infinity, ξ_0 and ξ_∞, by relationship[8]

$$\xi_t = \xi_0 + (\xi_0 - \xi_\infty)\exp(-t/\tau_r) \qquad (47)$$

If the spectral distribution is known, using the frequency of the center of gravity of the spectral distribution ν, one can obtain the equation

$$\frac{\nu - \nu_\infty}{\nu_0 - \nu_\infty} = \frac{\tau_r}{\tau_f + \tau_r} \qquad (48)$$

where v_0 and v_∞ are positions of the center of gravity of the spectral distribution when t = 0 and $t \to \infty$, respectively.

Several authors have obtained a general expression for the shift of electron transitions at is caused by the universal interactions. For example, the expression obtained by Bakhshiev[7] at $\tau_f >> \tau_r$ is

$$hc\Delta v^{a,f} = C_1^{a,f} \frac{2n^2+1}{n^2+2}\left(\frac{\varepsilon-1}{\varepsilon+2} - \frac{n^2-1}{n^2+2}\right) + C_2 \frac{2n^2+1}{n^2+2}\frac{n^2-1}{n^2+2} + \\ + C_3 \frac{n^2-1}{n^2+2} + C_4^{a,f} \frac{n^2-1}{2n^2+1} \qquad (49)$$

where $C_1^{a,f}$, C_2, C_3, and $C_4^{a,f}$ are parameters which depend mostly on the properties of the solute molecule and characterize a contribution to the spectral shift of the orientational, inductional, dispersional, and polarizational interactions; n and ε are refraction index and dielectric constant of the medium. Indices a and f refer to absorption and fluorescence, respectively.

The orientational interaction described by the first term in Equation 49 is the interaction which was mentioned above between the rigid dipole of the solute molecule and the constant dipoles of the solvent. Inductional interaction (the second term in Equation 49) is the interaction between the rigid dipole of the solute and the dipoles which it induces in molecules of the solvent, and, similarly, between constant dipoles of the solvent and an induced dipole in the solute molecule. The third term reflects dispersional interactions that are the interactions between the instantaneous dipole moments arising due to the electron movements in the solute molecule and the dipoles that are induced by them in the neighboring molecules. The polarization effect reflected in the fourth term is the interaction of the internal electric field of the solvent, with an oscillating dipole arising as a result of the absorption of a light quantum by the solute molecule.

For convenience in comparing experimental data with theoretical expressions, the latter can be represented in the following form:

$$hc\Delta v^{a,f} = C_1^{a,f} \cdot F^{a,f}(\varepsilon, n) \qquad (50)$$

If we have absorption and emission spectra, it is convenient to use Stocks shift changes instead of the shifts $\Delta\nu^{a,f}$:

$$\Delta\nu^{a-f} = \Delta\nu^{a} - \Delta\nu^{f} \tag{51}$$

$$hc\Delta\nu^{a-f} = \Delta C^{a-f} \cdot f(\varepsilon, n) \tag{52}$$

$$\Delta C^{a-f} = C_1^a - C_1^f \tag{53}$$

The functions $F^{a,f}(\varepsilon,n)$ and $f(\varepsilon,n)$ are called functions of universal interaction.

Bakhshiev showed that the method of universal interaction functions describes rather well the spectral shifts for many molecules of various structures and electron transitions of different types and also phenomena taking place in solvents of various kinds over a wide temperature range. Similar equations were also obtained by other authors.[10-13]

As a rule, it is very difficult to take into account specific interactions leading to formation of stoichiometric complexes of the solute molecule with the solvent molecules. Their energy is often higher than that of the universal interactions, and sometimes a change in the electron configuration of the solute molecule is so significant that it can be considered as an entirely different molecule. The most studied specific interactions are now probably hydrogen bonds and charge transfer complexes.

XIII. SCHEME OF STEADY-STATE SPECTROFLUORIMETER

At present there exist many different constructions of spectrofluorimeters, some of which are used in commercial production of these instruments. Nevertheless, the main scheme of these instruments is practically the same (Figure 12).

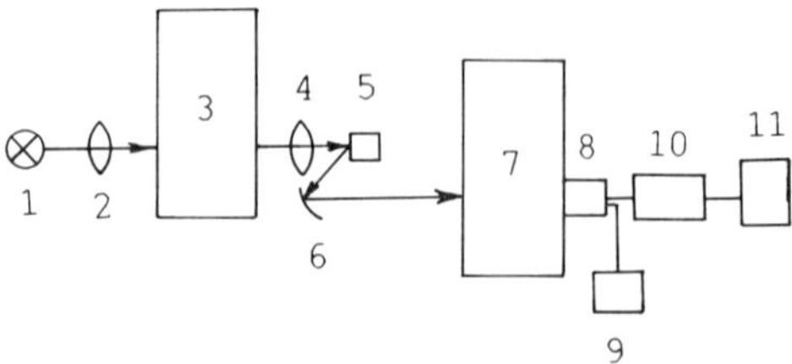

Figure 12. Scheme of a steady-state spectrofluorimeter: (1) excitation light source; (2) light-focusing arrangement; (3) exciting monochromator; (4) light-focusing arrangement; (5) measuring cell; (6) light-collecting and light-focusing arrangement; (7) analyzing monochromator; (8) photomultiplier; (9) high-voltage power source for the photomultiplier; (10) amplifier of electric signals; and (11) analyzing and recording devices.

Lamps with both continual and line emission spectra are used as a source of exciting light (Figure 12, 1). Lamps with a continuous emission spectrum (for example, hydrogen or deuterium lamps) are convenient in that they permit choice of any wavelength for excitation of fluorescence. Lamps with line spectra (for example, mercury lamps) give a more intense emission, but the choice of excitation wavelength is limited. The spectrum of a super high-pressure mercury lamp contains a set of lines suitable for selective excitation of different aromatic amino acid residues in proteins. For example, the light of the mercury line at 296.7 nm mostly excites fluorescence of tryptophan residues, while the light of the line at 280.4 nm excites both tyrosine and tryptophan residues. Mercury lines at 257.5 or 265 nm (265.2; 265.4; 265.5 nm) are convenient for excitation of phenylalanine residues.

The light from the excitation source is focused to the entrance slit of the exciting monochromator (Figure 12, 3), which separates the light with a narrow spectral distribution. Studies of protein luminescence require a rather high quality of the exciting monochromator: it must provide intense lines with narrow spectral distribution, i.e., it must have high power and high spectral resolution to be able to work with a spectral slit width of 1 to 2 nm.

The light from the exit slit is focused to the cell with a sample

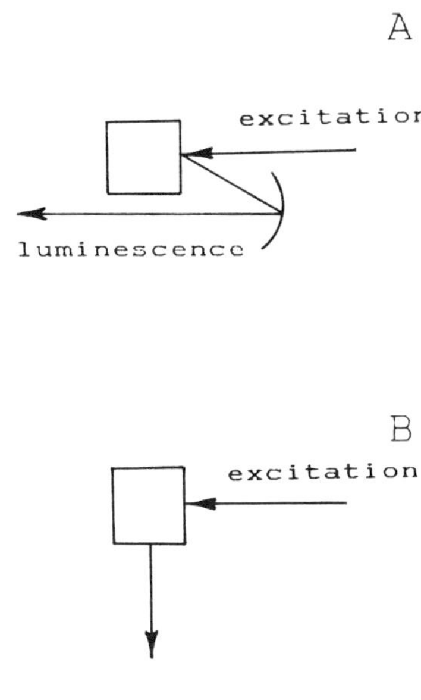

Figure 13. Two schemes for light collection from the cell with a sample. (A) From the front surface; (B) from the side surface.

(Figure 12, 5). At temperatures above 0°C, quartz cells with different optical path lengths are commonly used for the measurements. Fluorescence light can be collected either from the front or from the side surface of the cell (Figure 13). The second scheme is commonly used in most commercial instruments, while the scheme of the light collection from the front surface of the cell possesses some important advantages. First, this scheme allows measurement of luminescence not only of transparent liquid samples, but also of nontransparent liquid or even nontransparent solid samples. Second, such a scheme permits adequate correction of the emission spectra for reabsorption and screening effects.

If a protein solution contains a substance which absorbs light at the excitation wavelength, then a fraction of the light will be absorbed by this substance and not by protein chromophores, i.e., the substance will "screen" the protein. Besides that, if the solution contains a substance absorbing in the range of the protein emission, a fraction of the protein fluorescence will be absorbed by the substance (reabsorption). In the case of the collection of fluorescence light from the front surface of the measuring cell, the coefficient taking into account these effects can be calculated from the equation[14]

$$w(\lambda) = \frac{I_{true}}{I_{measured}} = \frac{1 - T_p}{1 - T_p T_e T_r(\lambda)} \cdot \frac{D_p + D_e + D_r(\lambda)}{D_p} \tag{54}$$

where T is transparency and D is absorbance ($T = 10^{-D}$, $D = \varepsilon \cdot l \cdot C$, ε is extinction coefficient, l is optical path length, C is concentration of absorbing substance). Indices p and e refer to the protein and screening agent at the excitation wavelength, while r refers to the reabsorbing agent at fluorescence wavelength. Figure 14 shows how to measure these values. The absence of proper correction of fluorescence spectra for reabsorption and screening effects may cause essential errors.

In order to carry out measurement at various temperatures, the cell holder is usually thermostatted.

In some types of spectrofluorimeters two cells are used. This allows one to work in the differential regime, i.e., to measure a difference between fluorescent signals obtained from two samples.

Fluorescence light collected from the sample is focused to the entrance slit of the analyzing monochromator (Figure 12, 7), which decomposes the fluorescence light into a spectrum. The monochromator for spectral studies of proteins should have high spectral resolution and high power. Spectral slit width should be no more than 1 to 2 nm and in some cases even less than 1 nm.

After monochromator, the light goes to the photocathode of the photomultiplier (Figure 12, 8), which changes the light signal to an electrical signal. The electrical signal is amplified by an amplifier (Figure 12, 10) and then analyzed and recorded or only recorded by a recorder (Figure 12, 11).

Open metal cells placed into a quartz Dewar vessel are used for low-temperature measurements. In order to prepare a sample, the protein solution is poured into previously cooled cell closed by a cover; the cover is removed after freezing of the sample.

Since each element of a spectrofluorimeter possess its own

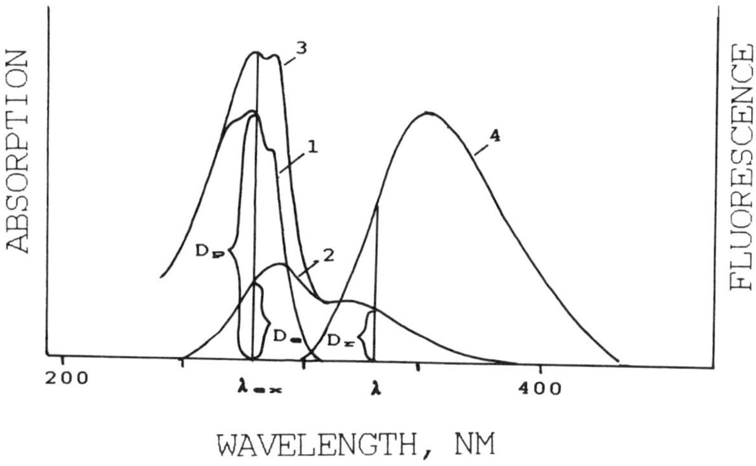

Figure 14. Absorption spectra of a protein (1) and a reabsorbing and screening substance (2). (3) Total absorption of the solution; (4) protein fluorescence spectrum.

spectral sensitivity, the measured spectrum is not the true one. In order to obtain the true emission spectrum, the measured spectrum should be corrected for spectral sensitivity of the instrument. This can be done either automatically during the course of the spectral measurements or in the course of further spectrum treatment. The simplest method of obtaining a spectral sensitivity curve is measurement of the emission spectrum of a substance with known true spectrum. The wavelength dependence of coefficient $k(\lambda) = I_{measured}(\lambda)/I_{true}(\lambda)$ is the spectral sensitivity curve. For studies of protein luminescence it is reasonable to use luminescence spectra of tryptophan, tyrosine, and phenylalanine as calibration spectra for obtaining the spectral sensitivity curve of an instrument.

At present, practically all the commercially available spectrofluorimeters are computerized. This simplifies the spectral measurements, but sometimes leads to some specific additional problems which are not discussed in this book.

There are several different ways to carry out luminescent kinetic measurements. Now we can measure fluorescence lifetime by means of phase-modulation, frequency-domain, or time-resolved methods. The instrumentation for such measurements is more sophisticated and more expensive in comparison with the equipment for steady-state studies. Since this book is devoted mostly to steady-state luminescence, the instrumentation for kinetic fluorescence measurements is not considered here.

REFERENCES

1. Landau, L. D. and Lifshitz, E. M., *Quantum Mechanics. Non-Relativistic Theory*, Physics and Mathematics Literature Publishing, Moscow, 1963.
2. Birks, J. B., *Photophysics of Aromatic Molecules*, John Wiley & Sons, New York, 1970.
3. Murrell, J., *Theory of Electronic Spectra of Organic Molecules*, Methuen, London, 1963.
4. Turro, N. J., *Molecular Photochemistry*, W. A. Benjamin, New York, 1965.
5. Parker, C. A., *Photoluminescence of Solutions with Applications to Photochemistry and Analytical Chemistry*, Elsevier, Amsterdam, 1968.
6. Calvert, J. G. and Pitts, J. N., Jr., *Photochemistry*, John Wiley & Sons, New York, 1966.
7. Bakhshiev, N. G., *Spectroscopy of Intermolecular Interactions*, Nauka, Leningrad, 1972.
8. Meyster, T. G., *Electronic Spectra of Multi-Atomic Molecules*, Leningrad University Publishing, Leningrad, 1969.
9. Förster, T., *Ann. Phys. (Leipzig)*, 2, 55, 1948.
10. Von Lippert, E., *Z. Electrochem.*, 61, 962, 1957.
11. Kawski, A., *Acta Phys. Pol.*, 29, 507, 1966.
12. Mataga, N., Kaifu, Y., and Koizumi, M., *Bull. Chem. Soc. Jpn.*, 29, 465, 1966.
13. Liptay, W., *Z. Naturforsch. Teil A*, 20A, 1441, 1965.
14. Burstein, E. A., *Biofizika*, 13, 433, 1968.

Chapter 3

Spectroscopic Properties of Isolated Protein Chromophores

I. INTRODUCTION

Absorption and emission of proteins in the near-UV region is due, first of all, to tryptophan, tyrosine, and phenylalanine residues, which are natural reporter groups in protein molecules (Figure 1). Interpretation of fluorescence spectra of proteins requires knowledge of spectroscopic properties of these amino acids in the isolated state.[1-7]

II. TRYPTOPHAN

A. Absorption Spectrum

Absorption and fluorescence spectra of tryptophan in aqueous solution are shown in Figure 2. Tryptophan possesses two absorption bands: one with maximum at 280 nm (ε_{280} = 5600 M^{-1} cm^{-1}) and another with maximum at 218 nm (ε_{218} = 33,000 M^{-1} cm^{-1}) (only one band is shown in Figure 2). The longer wavelength band has a weak vibrational structure: a shoulder at 271 nm, the main maximum at 279.5 nm, and a sharp peak at 288 nm. Absorption bands are caused by electron-vibrational $\pi \rightarrow \pi^*$ transitions in the indole ring, the aromatic π-system of which is formed by ten π-electrons. Quantum mechanical calculations show an uneven distribution of electron density among the atoms of the indole ring. This results in a large molecular dipole moment

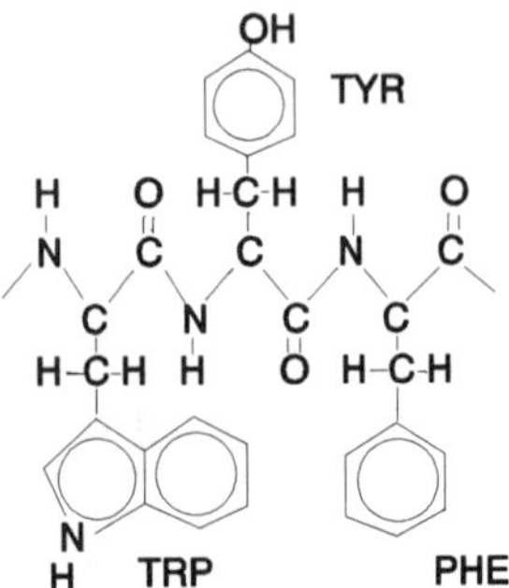

Figure 1. A fragment of polypeptide chain containing tryptophan, tyrosine, and phenylalanine residues.

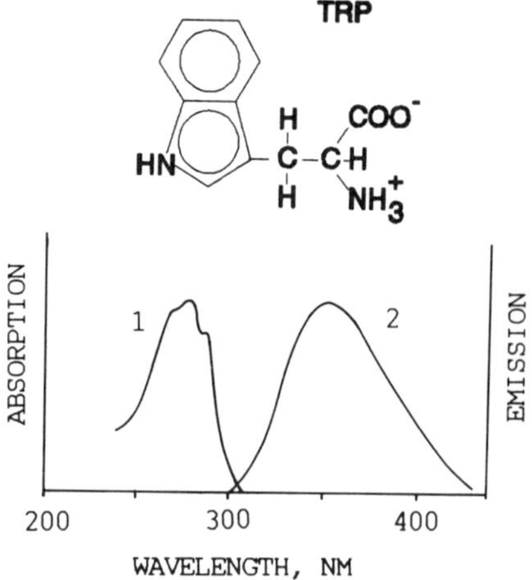

Figure 2. Absorption (1) and fluorescence (2) spectra of tryptophan in aqueous solution at neutral pH and room temperature.

and pronounced change of the dipole moment during the transition of the molecule to the excited state. It allows the possibility of dipole-dipole and specific interactions between the chromophore and polar solvent molecules.

The longer wavelength band of the indole chromophore in the region between 250 to 290 nm is assumed to be composed of two electron transitions. Analysis of the polarization fluorescence spectra revealed two superimposed absorption bands with almost perpendicular transition vectors which were ascribed to $^1L_a \leftarrow A$ (transition vector along the short axis) and $^1L_b \leftarrow A$ (transition vector along the long axis) transitions. The absorption band with the maximum at 218 nm contains one more electron transition $^1B_b \leftarrow A$.

An increase in solvent polarity causes a shift of the tryptophan absorption spectrum towards longer wavelengths. The shift for ^1L_a is more pronounced than that for ^1L_b. The shift is induced by both universal and specific interactions due to the formation of complexes with solvent molecules with participation of the NH-group of the indole ring. It should be noted, however, that the changes in the absorption spectrum of the indole chromophore induced by the solvent polarity changes are not very large, so that in order to detect them, one should use differential spectrophotometry methods. Similar changes in the fluorescence spectrum are much more pronounced.

B. **Emission Spectra**

In neutral aqueous solution at room temperature tryptophan displays a wide, structureless fluorescence band with maximum at 353 nm and a width of about 60 nm (Figure 2). It is worth noting that tryptophan fluorescence spectra measured in the same conditions but by different instruments may differ in both maximum position and shape, depending on the calibration spectra used. Unfortunately, at present there are no generally accepted standards for the spectral calibration of instruments in the near-UV region. Table 1 contains normalized fluorescence spectra of tryptophan, tyrosine, and phenylalanine in water at neutral pH which we use for the calibration of our spectrofluorimeters.

The position of the tryptophan fluorescence spectrum depends slightly on the ionic state of carboxyl and amino groups in the alanine side chain: at acidic pH values (cationic form) the tryptophan fluorescence spectrum is blue shifted, while at alkaline pH values (anionic form) it is red shifted in comparison with that of the zwitterionic form. These shifts of the fluorescence spectrum are always accompanied by similar shifts of the absorption spectrum.

The tryptophan fluorescence spectrum is extremely blue shifted (the maximum at 310 nm) in nonpolar solvent. Addition of polar compounds results in a red shift of the spectrum. The red shift of the fluorescence spectra of indole and its derivatives induced by the change in solvent polarity is thought to be caused by relax-

Table 1

Normalized Fluorescence Spectra of Water Solutions of Tryptophan, Tyrosine, and Phenylalanine at Neutral pH and Room Temperature

λ (nm)	Trp	Tyr	λ (nm)	Phe
280		21	265.00	90
285		193	266.25	171
290		565	267.50	260
295		870	268.75	350
300	11	986	270.00	453
305	35	981	271.25	550
310	95	865	272.50	707
315	190	765	273.75	779
320	315	541	275.00	806
325	473	400	276.25	800
330	623	292	277.50	827
335	772	211	278.75	927
340	880	149	280.00	997
345	957	104	281.25	982
350	996	74	282.50	936
355	996	52	283.75	873
360	949	38	285.00	815
365	886	27	286.25	788
370	807	20	287.50	779
375	718		288.75	738
380	633		290.00	722
385	549		295.00	522
390	468		300.00	378
395	395		305	262
400	335		310	180
410	242		315	126
420	176		320	86
430	130		325	63
440	98		330	47
450	73			

ation processes of the orientational interactions of the chromophore dipole with the solvent dipoles. Specific interactions between the solvent molecules and the chromophore in the ground and excited (exciplexes) states also contribute significantly to the shift. The excitation causes such pronounced changes in the electron density distribution in the indole ring that its pK_a of protonation and deprotonation changes by 6 to 8 units, making the

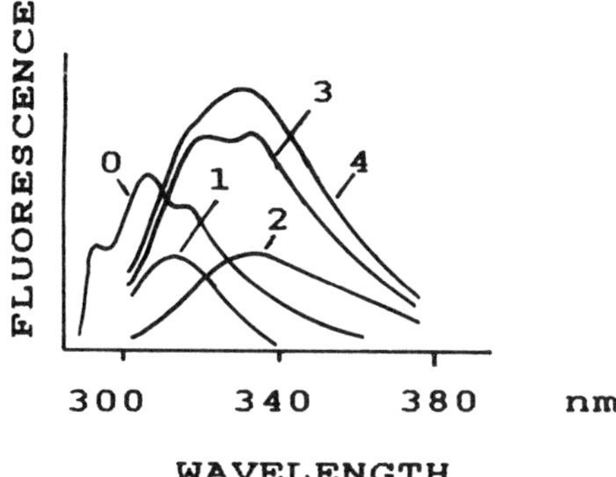

Figure 3. Fluorescence spectra of 3-methylindole in *n*-heptane (0) and its exciplexes 1:1 (1) and 2:1 (2) with *n*-butanol, and total spectra of the exciplexes in *n*-heptane in the presence of 11.2 m*M* (3) and 14.8 m*M* (4) *n*-butanol. (From Burstein, E. A., *Mol. Biol. (U.S.S.R.)*, 17, 455, 1983. With permission.)

>NH group of the ring a good donor and the C(3) atom a good acceptor of a proton upon formation of hydrogen bonds. As a result, proton acceptors, for instance, dioxane and ethers, form 1:1 exciplexes with indole, while the hydroxyl-containing molecules form with indole 2:1 exciplexes. A large energy of exciplex formation results in pronounced spectral shifts (Figure 3).

The formation of exciplexes often makes the ^1L_a level lower than the ^1L_b one, and the emission proceeds mostly from the ^1L_a level in this case. It should be noted that both dipole relaxation and exciplex formation require nanosecond mobility of the molecules in the indole chromophore environment.

The freezing of an aqueous solution of tryptophan down to –196°C immobilizes water molecules, but only crystals of tryptophan (fluorescence maximum at 340 nm) emit in the pure ice. The tryptophan crystals are formed upon the freezing of the aqueous solution of tryptophan. In the course of freezing tryptophan in aqueous solution in the presence of electrolytes, some of the tryptophan molecules come into contact with the electrolyte ions and begin not only to fluoresce with a maximum at 325 nm, but also to phosphoresce (Figure 4). The phosphorescence spectrum of tryptophan in the frozen water-salt solution has three characteristic maxima at 406, 432, and 456 nm.

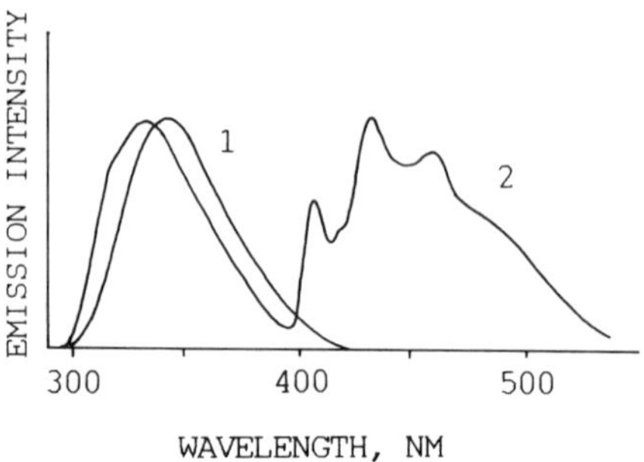

Figure 4. Emission spectra of frozen aqueous solutions of tryptophan in water (1) and in 1 *M* NaCl (2) at –196°C.

C. **Emission Lifetime**

The fluorescence decay curve for tryptophan and other indole derivatives in water at room temperature cannot be approximated by a single exponent. It can be fitted only by two exponents, with their relative contributions dependent on wavelength.[9] The lifetimes of the components are 3.1 and 0.51 ns. Their emission spectra display maxima at 350 and 330 nm, respectively. The presence of the two components in the emission of tryptophan in water seems to reflect the existence of two tryptophan conformers differing in the orientation of the alanyl part. It is not excluded, however, that this might be explained by a fundamental nonexponentiality of the tryptophan fluorescence decay.

The phosphorescence decay curve for the water-salt solution of tryptophan at –196°C is fitted by a single exponential. In the case of salts which do not affect the spin-orbit coupling, the phosphorescence lifetime of tryptophan is about 5.5 s.

D. **Analytical Description of the Fluoresence Spectrum Shape**

It is clearly seen from Figure 2 that the fluorescence spectrum of an aqueous solution of tryptophan is asymmetric and therefore cannot be approximated adequately by either a Gaussian or Lorenztian distribution. It turns out that if the emission spectrum of tryptophan and other indole derivatives is plotted on the

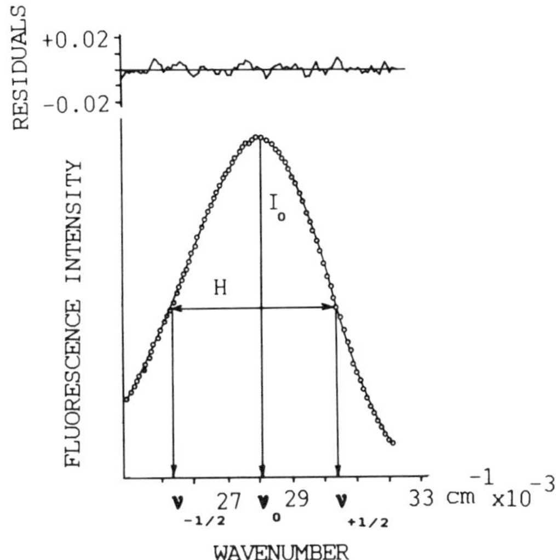

Figure 5. Approximation of fluorescence spectrum of tryptophan in water by lognormal distribution. The points are experimental, and the curve is a theoretical one computed according to Equation 1 and fit to the experimental points by variation of ν_0, $\nu_{+1/2}$, and $\nu_{-1/2}$. (Upper panel) Distribution of residuals. (The data were kindly supplied by Dr. Emelyanenko.)

frequency scale, it can be correctly approximated by the log-normal distribution (Figure 5):[1,2]

$$I(\nu) = \frac{I_0 b}{a-\nu}\exp(-C^2/2)\cdot\exp\left\{-\frac{1}{C^2}\left[\ln\left(\frac{a-\nu}{b}\right)\right]^2\right\} \tag{1}$$

$$I(\nu) = 0 \; \textit{if}\; \nu < a$$

where I_0 is maximal amplitude, ν_0 is spectral maximum position, H is spectrum width, and ρ is spectral asymmetry:

$$\rho = \frac{\nu_{+1/2} - \nu_0}{\nu_0 - \nu_{-1/2}} \tag{2}$$

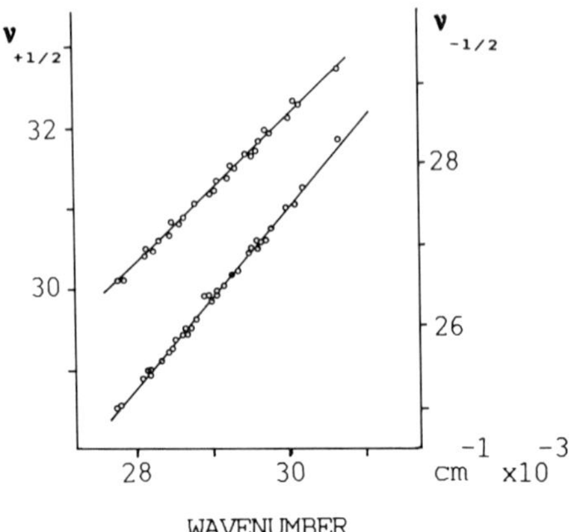

Figure 6. Dependence of $\nu_{+1/2}$ and $\nu_{-1/2}$ on ν_0 for 19 indole derivatives in various conditions. (The data were kindly supplied by Dr. Emelyanenko.)

where $\nu_{\pm 1/2}$ are positions of the half-amplitude points.

$$a = \nu_0 + H\frac{\rho}{\rho^2 - 1} \tag{3}$$

$$b = H\frac{\rho}{\rho^2 - 1}\exp(-C^2) \tag{4}$$

$$c = \frac{\ln \rho}{(2\ln 2)^{1/2}} \tag{5}$$

The distribution contains four independent parameters: I_0, ν_0, $\nu_{+1/2}$, and $\nu_{-1/2}$. Studies of fluorescence spectra of indole derivatives in various conditions have shown that there is a linear relationship between ν_0 and $\nu_{\pm 1/2}$ (Figure 6). This allows description of a spectrum by only two parameters I_0 and ν_0.

E. **Fluorescence Quantum Yield**

The data on absolute fluorescence quantum yield of tryptophan in water are contradictory. Different authors report different values of the fluorescence yield: from 0.11 to 0.20 at pH 7 and 20°C. For this reason it is better to use relative fluorescence quantum yield.

Deactivation of the excited state of the indole chromophore has been extensively studied by many authors,[1,3] and some peculiarities of the processes are now well known. The most common method used in these studies is the study of the effects of D_2O substitution for H_2O, addition of various substances, and pH and temperature changes on fluorescence yield value.

At room temperatures the substitution of H_2O by D_2O causes an increase in the fluorescence quantum yield value for indole and its derivatives, with almost no shifts of the absorption and fluorescence spectra. For tryptophan at neutral pH values, fluorescence quantum yield increases by a factor of 2.1. The observed isotopic effect is caused, first of all, by the deuteration of water as a solvent, but not by an isotopic substitution in the chromophore. The D_2O effect decreases upon protonation or substitution of the amino group in the side chain of tryptophan (1.06- to 1.54-fold increase of fluorescence yield), which is explained by a pronounced isotopic effect for the intramolecular fluorescence quenching by the amino group.

Studies of lowering of the isotopic effect by electrolytes and low-temperature measurements in H_2O and D_2O allow one to conclude that the isotopic effect of D_2O for indole derivatives at room temperature is caused by kinetic quenching interactions of water with the excited chromophore. Intersystem crossing to the triplet state does not take part in these processes. Since different deactivation processes can depend differently on temperature, the common method of study of these is analysis of the temperature dependence of fluorescence quantum yield. In such an analysis, the rate constant of the emission transition, k_f, is assumed to be independent of temperature, while the rate constant of the $S_1 \rightsquigarrow T_1$ intersystem crossing is thought to be only slightly dependent on temperature. There are reasons to believe that the process $S_1 \rightsquigarrow S_0$ has very low probability; therefore, it is usually neglected in the analysis.

Temperature dependence of fluorescence quantum yield of indole derivatives is described by the equation

$$q = \frac{k_f}{k_f + k_2' + k_3^{0\prime} \exp(E/RT)} \qquad (6)$$

or

$$1/q = 1 + k_2 + k_3^0 \exp(E/RT) \qquad (7)$$

where k_f is the rate constant of the emission transition and k_2' and $k_3^{0\prime}$ are the rate constants of the nonradiative transitions. Strictly speaking, Equations 6 and 7 are applicable only to situations in which the chromophore has no quenching groups in the side chain; however, even when they are present, the parameters calculated with these equations usually are not senseless. For indolyl acetic and indolyl propionic acids, and for indole which contains no quenching groups in the side chain, k_2 lies within the limits from 0.1 to 0.4, and E is within the region from 12.2 to 13.4 kcal/mol. The process described by the k_3 constant gives the main contribution to the deactivation of excitation in aqueous solutions. It is just this constant that decreases 2.3 to 2.6 times in D_2O.

The analysis of the thermal quenching curves for indole derivatives with quenching groups in the side chain can be carried out, assuming that the introduction of a new quenching group into the molecule does not change the rate constants of other processes. This means, for example, that for tryptophan, k_2 and k_3 do not differ from those for indolyl propionic acid, but that there appears an additional quenching by the amino group with an activation energy of 4.7 kcal/mol.

The thermally activated, nonradiative deactivation of excited states of the indole chromophore is thought to be caused by a photochemical quenching interaction of the excited indole ring with water molecules. The fact that k_3 decreases without any change in E, which is induced by gradual addition of dioxane or various electrolytes to the aqueous solution of indole, is in favor of this quenching mechanism. Calculations show that proton transfer in neutral aqueous solution cannot result in the observed high rate constants. The thermally activated, nonradiative deactivation of the indole excited state is caused by photoionization of the indole ring, i.e., by transfer of an electron from the excited chromophore to water. The excitation of the indole chromophore induces redistribution of charges in the molecule, resulting in localization of high electron density on one of its atoms. There exists a possibility that during the excitation lifetime a special arrangement from several water molecules (3 to 4) will be formed which by itself or in combination with the indole ring will create a potential hole that can trap an electron from the excited ring. It is just the creation of such an arrangement which requires the activation energy. The electron transfer disturbs the aromaticity of the chromophore; therefore, it loses the excitation energy to thermal vibrations. The resulting electron-deficient radical rapidly recombines with the electron, which gives the chromophore in the ground state. It is possible, however, that the complete electron transfer does not proceed and that the deactivation is achieved at the expense of a trap-induced internal conversion.

The temperature-independent constant k_2 for indole derivatives in aqueous solution is 2.5 to 3.0 times higher than the constant k_3. The substitution of H_2O by D_2O and addition of nonquenching electrolytes do not influence its value. Intersystem crossing to the triplet level is assumed to give the main contribution to the k_2.

F. **Acid-Base Properties in the Ground and Excited States**

The curves of spectrophotometric and spectrofluorimetric pH titrations for the indole chromophore are essentially different. Attachment of a proton to the indole ring proceeds at acidic pH values in the ground state with the midpoint at 9.4 *M* H_2SO_4. Dissociation of a proton from the amino group of the indole ring in the ground state occurs only at extremely alkaline pH values. On the other hand, quenching of the indole chromophore takes place at pH values below 1.7 and above 12.0, where as yet nothing proceeds in the ground state.

Deprotonation of the indole ring in the excited state accompanied by fluorescence quenching proceeds within the pH region from 11 to 14. Tryptophan with an ionized imino group displays weak fluorescence centered at 420 to 430 nm. The Stern-

Volmer constant for the quenching by OH^- ions is 63 M^{-1}, which corresponds to the quenching rate constant 0.8×10^{10} M^{-1} cm^{-1}.

G. **Fluorescence Quenching by Protein Groups**

Almost all polar protein groups are able, to some extent, to quenching the fluorescence of the indole chromophore. The mechanism of this quenching in some cases is still far from clear.

1. Carboxylate (–COO⁻) and Carboxyl (–COOH) Groups

Aspartic and glutamic acid residues in proteins contain carboxylic groups. pK values of their protonation lie in the region from 4 to 5. pK of the C terminal carboxyl is 2.5 to 3.5. At pH > pK the $-COO^-$ group almost does not quench the fluorescence of the indole chromophore. In contrast to this, neutral carboxyl effectively quenches indole fluorescence, with the quenching having a dynamic character. The lower the pK_a value of the carboxyl, the stronger are its quenching properties. The protonation of the tryptophan carboxylate increases the total constant of the nonradiative deactivation by a factor of 2.5.

It is assumed that carboxyl quenches the fluorescence of the indole chromophore by means of a proton transfer to the excited ring. The direct quenching by protons seems to compete with the quenching by water, accompanied by formation of solvated electrons which are able to interact with protons.

2. Amide (–CONH$_2$) and Peptide (–CO–NH–) Groups

Both amide and peptide groups in the side chain of the indole chromophore strongly quench its fluorescence. External amide and peptide groups also quench indole fluorescence, their quenching action requiring water. Since these groups are active electron scavengers, it is reasonable to assume that they quench fluorescence by means of accepting of electron. Taking into account the essential role of water in this process, one can believe that these groups act not directly, but by arranging water molecules in the hydration shell in such a way that the activation energy for creation of the quenching electron trap decreases.

3. Ammonium (–NH_3) and Amino (–NH_2) Groups

ε-Groups of lysines and N terminal α-groups of proteins are amino groups. Ammonium group of the side chain of the indole chromophore quenches its fluorescence with the process being dynamic. Deprotonation of the ammonium group increases indole fluorescence quantum yield. External ammonium groups also quench indole fluorescence. The quenching effect essentially increases with the decrease in pK_a of the ammonium group. The quenching is thought to proceed by electron transfer from the excited chromophore to ammonium.

–NH_2 groups also quench indole fluorescence, and their quenching effect increases with the increase in their pK_a values. In this case, the quenching is achieved by protolysis of the indole ring enamine by the amino group.

4. Guanidine Group (–NH–C(NH_3)(NH_3))

Only arginine residues in proteins have the guanidine group It is known that arginine is able to quench fluorescence of the indole chromophore, but the mechanism of this process is to be elucidated.

5. Imidazole (N NH) and Imidazolium (HN + NH) of Histidine

Histidine at pH <6 to 7 is an effective quencher of fluorescence of the indole chromophore. The quenching occurs as a result of formation of complexes between the indole ring and histidine by means of the stacking interaction, which is well seen from the proton nuclear magnetic resonance (NMR) spectra.

The nonprotonated form of imidazole is also able to quench indole fluorescence, although the efficiency of the quenching is 14 to 15 times lower than that of imidazolium and is explained by dynamic effects. The most probable mechanism of the fluorescence quenching in this case is a proton transfer from the excited chromophore to imidazolium.

6. Sulfur-Containing Groups (–SH, –S⁻, –S–S–, –C–S–C–)

The thiol group effectively quenches indole fluorescence. The Stern-Volmer constant for the quenching of indole fluorescence by β-mercaptoethanol in ethanol is 11 M^{-1}. Mechanism of the quenching is not quite clear now, though it was suggested that the quenching action of thiol groups is due to electron transfer from the excited indole chromophore.

A more efficient quencher is the thiolate group ($-S^-$). The Stern-Volmer constant for the quenching of indole fluorescence in water by β-mercaptoethanol is 14 M^{-1}. In this case the quenching is thought to be due to the proton transfer from the excited chromophore to the quencher.

Disulfide is one of the strongest quenchers of indole fluorescence. The Stern-Volmer constant for the quenching of indole fluorescence by the compound $CH_3–CH_2–S–S–CH_2–CH_3$ is 38 M^{-1}. The quenching process in this case seems to be due to either a facilitation of the intersystem crossing to the triplet state or a photoionization of the chromophore.

H. Temperature Dependence of Bimolecular Quenching of Indole Fluorescence

The expression for the quantum yield of indole fluorescence taking into account additional deactivation processes occurring due to bimolecular interaction of the chromophore with quencher molecules is[3]

$$a/q = 1 + k_2 + \alpha(c) \cdot k_3^0 \cdot \exp(-E/RT) + k_i(T) \cdot c \qquad (8)$$

where a is the fraction of the fluorescent molecules which do not form a nonemitting complex:

$$a = 1/(1 + K \cdot c) \qquad (9)$$

(K is the constant of the complex formation, c is quencher concentration) α(c) is a coefficient characterizing the effects of the quencher on the constant of the quenching by water, k_i is a rate constant of kinetic quenching by the added molecules.

Compounds which form a nonfluorescent complex with indole chromophore are, for example, imidazole derivatives and acrylamide commonly used as a neutral quencher in studies of proteins. The temperature dependence of this process can vary depending on the nature of the forces governing the complex formation.

The efficiency of the kinetic quenching (k_i) of indole fluorescence displays similar temperature dependence over a wide range of quenchers (imidazolium, sulfhydryl, carboxyl, ammonium groups, H^+, I^-, NO_3^-, and so on). The curve does not depend on the quenching mechanism or on the absolute value of the constant. Exceptions are the processes of quenching by proton acceptors which are accompanied by an appearance of fluorescence of indolate. The universal temperature dependence of k_i cannot be described by a strict Arrhenius equation, but is in crude approximate correspondence to it with activation energy of about 2 to 3 kcal/mol. The universal curve is not as steep as the temperature dependence of T/η (T is temperature and η is solvent viscosity). The universal character of the curve $k_i(T)$ shows that it reflects some common properties of the systems studied. The main factor for this curve is assumed to be the activation of diffusion. Its difference from the T/η curve is explained by nonproportionality of changes of quantum yield and lifetime values.

III. TYROSINE

A. Absorption Spectrum

Figure 7 shows the absorption spectrum of tyrosine in neutral aqueous solution (only the longer wavelength band is shown). Maxima of the spectrum are located at 222 nm (ε_{222} = 8000 M^{-1} cm^{-1}) and 275 nm (ε_{275} = 1230 M^{-1} cm^{-1}). The longer wavelength band possesses a poorly resolved vibrational structure with shoulders at 267 and 282 nm. The absorption is caused by $\pi \rightarrow \pi^*$ transitions. Deprotonation of the hydroxyl group of tyrosine essentially changes its absorption spectrum: it shifts towards longer wavelengths and its intensity increases. In this case the spectral maxima are at 240 nm (ε_{240} = 11,700 M^{-1} cm^{-1})

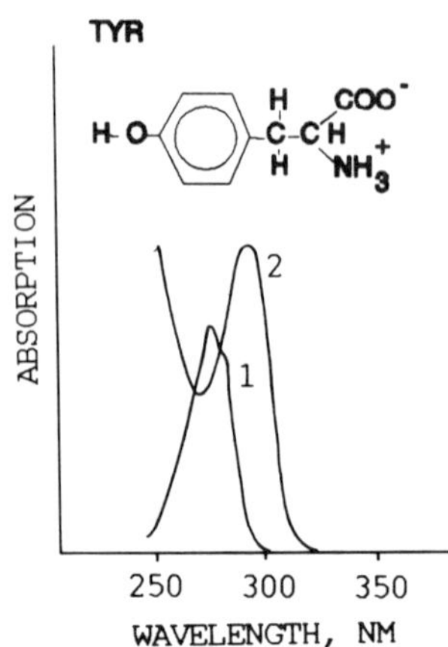

Figure 7. Absorption spectra of aqueous solution of tyrosine at pH 7 (1) and pH 12 (2).

and 293 nm (ε_{293} = 2340 M^{-1} cm^{-1}). pK_a of tyrosine ionization in water is 10.04 ± 0.03. Values of pK_a of ionization of tyrosine residues in proteins vary over a wide range depending on their environment.

B. Emission Spectrum

Figure 8 shows fluorescence spectra of tyrosine in water at neutral pH and room temperature. The maximum of the spectra is at 303 to 304 nm, and its width is 34 nm. The fluorescence yield of tyrosine in water is 0.21 at 20°C.[10]

Increasing the pH of the solution does not change the tyrosine spectral position, but decreases tyrosine fluorescence quantum yield with an apparent pK of 9.7. Acidification of the solution causes a similar effect with an apparent pK of 2.45, which corresponds to the pK of ionization of the carboxyl group.

Freezing of tyrosine solution shifts its fluorescence spectrum down to 298 to 299 nm, i.e., by 4 to 6 nm. Phosphorescence of tyrosine (Figure 8), with the maximum at 377 nm and shoulders at 400 to 410 nm, appears in frozen water-salt solutions at liquid nitrogen temperature.

The data presented demonstrate that in contrast to tryptophan, the position and shape of the emission spectrum of tyrosine

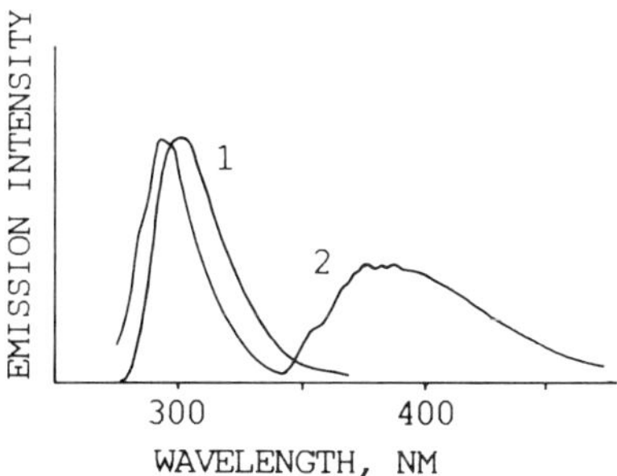

Figure 8. Emission spectra of tyrosine in aqueous solution at room temperature (1) and in 1 *M* NaCl at –196°C (2).

changes only slightly depending on conditions. This means that the tyrosine fluorescence contains less information than the tryptophan fluorescence. The informative parameters of tyrosine fluorescence are mainly the fluorescence yield and fluorescence lifetime. It should be noted that the fluorescence spectrum of aqueous solution of tyrosine is also described by the lognormal distribution.

C. Emission Lifetime

According to the data of time-resolved fluorescence spectroscopy, fluorescence lifetime of tyrosine in neutral aqueous solution at room temperature is 3.5 ± 0.2 ns. Phosphorescence lifetimes of tyrosine at neutral, acidic, and alkaline pH values are about 3, 3, and 0.9 s, respectively.

D. Fluorescence Quenching by Protein Groups

1. Carboxylate (–COO⁻) and Carboxyl (–COOH) Groups

Addition of carboxy acids, and also glutamic and aspartic acids, to neutral aqueous solutions of tyrosine essentially quenches its fluorescence. The mechanism of this process is thought to be a proton transfer from the hydroxyl group of the excited chromophore to the solvent. It is facilitated by a lowering of the pK of ionization of the tyrosine hydroxyl of the excited chromophore down to 3.5 to 4.0. It is also facilitated by the existence of hydrogen bonding between tyrosine and water molecules, even in the ground state. Neutral carboxyl is also a powerful quencher of the tyrosine fluorescence.

2. Amide (–CONH$_2$) and Peptide (–CO–NH–) Groups

Both these groups quench the tyrosine fluorescence. Amidation of the carboxyl of tyrosine decreases its fluorescence yield by a factor of two. Very low fluorescence yield is observed for unstructured polypeptides when the only quenching group is the peptide bond. Amide and peptide groups are able to form weak complexes with the phenolic group of tyrosine. A decrease in the medium polarity increases the complex formation. The quenching action of amide and peptide groups is explained by an electron transfer from the excited chromophore to these groups.

3. Ammonium (–NH$_3^+$) and Amino (–NH$_2$) Groups

The ammonium group can be a rather effective quencher of the phenolic fluorescence. The α-ammonium group and ε-NH$_3^+$ group of lysine are effective quenchers of tyrosine fluorescence. At the same time, deprotonation of the –NH$_3^+$ group of tyrosine does not change its fluorescence.

The amino group is also a quencher of tyrosine fluorescence. For example, the ε-NH$_2$ group of lysine is a good quencher of tyrosine fluorescence. It is assumed that in this case the quenching is caused by a proton transfer from the OH group of the excited phenol to the NH$_2$ group.

4. Guanidine Group (–NH<NH_3^+, NH_3^+>)

The guanidine group does not seem to be a quencher of the phenolic fluorescence.

5. Imidazole and Imidazolium of Histidine

At pH <6 to 7, histidine is an effective quencher of tyrosine fluorescence. The lowering of pH from 8 to 4 decreases fluorescence quantum yield of tyrosine in the presence of histidine by a factor of 1.5.

6. Sulfur-Containing Groups

Thiol and especially disulfide groups, are strong quenchers of phenolic fluorescence. The Stern-Volmer constants for fluorescence quenching of n-cresol in ethanol by β-mercaptoethanol and by CH_3–CH_2–S–S–CH_2–CH_3 are 3 and 45 M^{-1}, respectively.

It is clearly seen from the data presented that there exists little information concerning the fluorescence quenching of tyrosine fluorescence by protein groups.

IV. PHENYLALANINE

A. Absorption Spectrum

Figure 9 shows the absorption spectrum of phenylalanine in neutral aqueous solution at room temperature. The spectrum possesses a rich vibrational structure: maxima at 187.5, 205, 242, 252, 257 (major peak), 263, and 267 nm. Its molar extinction coefficient at 257 nm is about 200 M^{-1} cm^{-1}. Changes of pH or solvent polarity cause very small changes in the phenylalanine absorption spectrum.

B. Emission Spectrum

Figure 9 shows the fluorescence spectrum of phenylalanine in neutral aqueous solution at room temperature (curve 2). The spectrum possesses a distinct vibrational fine structure with maxima at 275, 282 (major peak), 289 nm, and a shoulder at 303 to 305 nm. At –196°C the shape of the phenylalanine spectrum is practically the same as that at room temperature.

The presence of the vibrational structure in the absorption and fluorescence bands at room temperature suggests that the dipole moment of the phenylalanine chromophore equals zero both in the ground and excited states.

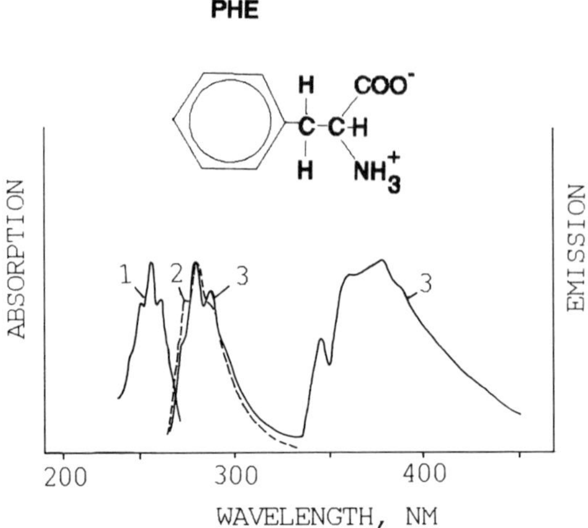

Figure 9. Absorption (1) and emission spectra of phenylalanine in neutral aqueous solution at room temperature (2) and in 1 *M* NaCl at –196°C (3).

At high concentrations the benzene chromophores of phenylalanine are able to form luminescent dimers in excited states, i.e., excimers, that display an emission centered at 320 nm.

The phosphorescence spectrum of phenylalanine at –196°C is shown in Figure 9 (curve 3).

C. Fluorescence Quantum Yield

The fluorescence quantum yield for phenylalanine in water is 0.038 to 0.045, i.e., essentially lower than those for tryptophan and tyrosine.[10] At acidic pH values, almost complete quenching of phenylalanine fluorescence occurs, which seems to be caused by neutralization of the carboxyl group. Alkaline pH values partially quench phenylalanine fluorescence, and this is accompanied by a 5-nm red spectral shift.

At present little is known about the quenching action of protein groups on phenylalanine fluorescence. It is known, for instance, that both external and side chain carboxyl groups at pH $> pK_a$ do not quench the fluorescence of the benzene chromophore. In contrast to this, neutral carboxyl is a powerful quencher of benzene fluorescence. It is also known that an ammonium group included in the side chain of phenylalanine or external ammonium will severely inhibit fluorescence quenching. Neutral amino group, guanidine group, and histidine quench benzene fluorescence.

It is evident from the data presented that the spectral properties of tryptophan have been most studied to date. This is explained by the fact that tryptophan fluorescence is most sensitive to changes in chromophore environment; therefore, researchers have studied its spectral properties most intensively.

REFERENCES

1. Burstein, E. A., *Luminescence of Protein Chromophores,* Vol. 6, Model Studies. Science and Technology Results. Biophysics, VINITI, Moscow, 1976.
2. Burstein, E. A., *Intrinsic Protein Luminescence,* Vol. 7, Origin and Applications. Science and Technology Results. Biophysics, VINITI, Moscow, 1977.
3. Busel, E. P., *Molecular Biology,* Vol. 3, Physical Methods in Molecular Biology, VINITI, Moscow, 1974, 85.
4. Chernitski, E. A., *Luminescence and Structural Lability of Proteins in Solution and in Cell,* Nauka i Tekhnika, Minsk, 1972.
5. Konev, S. V., *Fluorescence and Phosphorescence of Proteins and Nucleic Acids,* Plenum Press, New York, 1965.
6. Longworth, J. W., in *Excited States of Proteins and Nucleic Acids,* Steiner, R. F. and Weinryb, I., Eds., Pergamon Press, New York, 1971.
7. Demchenko, A. P., *Ultraviolet Spectroscopy of Proteins,* Springer-Verlag, Berlin, 1986.
8. Burstein, E. A., *Mol. Biol. (U.S.S.R.),* 17, 455, 1983.
9. Szabo, A. G. and Tayner, D. M., *J. Am. Chem. Soc.,* 102, 554, 1980.
10. Teale, F. W. J. and Weber, G., *Biochem. J.,* 65, 467, 1957.

Chapter 4

Protein Luminescence

I. INTRODUCTION

According to the Teale spectral classification,[1] proteins are divided into three classes: class A, proteins containing tyrosine residues, but not containing tryptophan residues; class B, proteins containing both tyrosine and tryptophan residues; and class C, proteins containing only phenylalanine residues. As a rule, proteins of class A display luminescence spectra which coincide with that of free tyrosine. Spectra of class B proteins are mainly due to the emission of tryptophan residues. The tyrosine component is revealed in this case only by means of a special analysis of the spectra. The small contribution of tyrosine emission in the fluorescence spectra of class B proteins is explained by the relatively low extinction coefficient of tyrosine, its low fluorescence quantum yield value, and the existence in some cases of excitation energy transfer from the tyrosine to tryptophan residues. Analogously, the absence of the phenylalanine contribution to the spectra of class A proteins is explained by the low extinction coefficient of phenylalanine and the existence of excitation energy transfer from phenylalanine to tyrosine chromophores. The fluorescence quantum yield for phenylalanine in proteins may be essentially higher than that of the free amino acid.

II. POSITION AND SHAPE OF PROTEIN FLUORESCENCE SPECTRA

A. Tryptophan Fluorescence

The position of the maximum of the fluorescence spectrum of tryptophan residues in proteins varies within the limits of 307 to

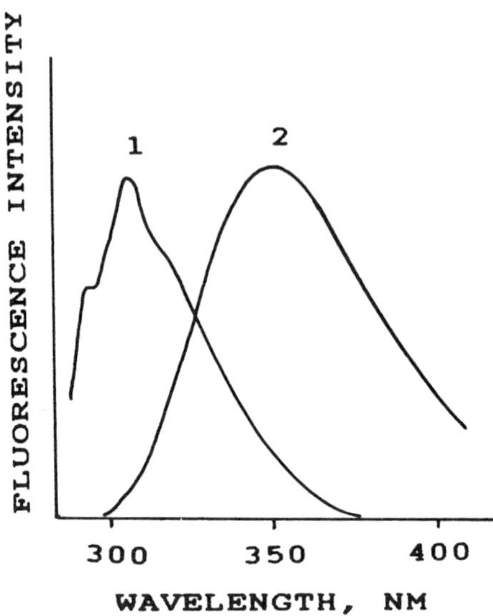

Figure 1. Fluorescence spectra of native azurin (1) and whiting parvalbumin denatured by 8 *M* urea (2).

353 nm. The shortest wavelength tryptophan fluorescence spectrum known at present is the emission spectrum of native azurin[2] (the main maximum at 307 nm), while fully unfolded proteins display the longest wavelength tryptophan fluorescence spectra (maximum at 353 nm as for free tryptophan in water) (Figure 1).

An analysis of the position and shape of tryptophan fluorescence spectra of a great number of proteins in various conditions and the data for fluorescence quenching by external quenchers allowed Burstein et al.[3-5] to formulate a model of discrete states of tryptophan residues in proteins. This model states that there exist at least five of the most probable spectral forms of tryptophan residues.

Spectral form A corresponds to the emission of the unperturbed indole chromophore in the neutral hydrophobic environment inside the protein globule. Such an emission was found in only two proteins: azurin and bacteriorhodopsin. Azurin, a small globular metal-binding protein from *Pseudomonas fluorescence,* contains a single tryptophan residue which displays a rather unusual fluorescence spectrum (Figure 2).[2] First, even at room temperatures it possesses a distinct vibrational structure. Second, its main emission maximum is located at 306.5 nm. Its second maximum is at 292.5 nm. The fluorescence and absorption spectra of azurin are characterized by a rather well-expressed mirror symmetry. Frequency analysis reveals nearly the same resolution of the main peaks and shoulders of the fine structure in the fluorescence spectra at room and low temperatures and also in the protein phosphorescence spectrum and the phosphorescence spectrum of free tryptophan in frozen water-salt solution (about 1200 to 1400 cm^{-1}). The freezing of azurin solution does not change the position of its fluorescence spectrum.

An azurin-like component is present also in fluorescence spectrum of bacteriorhodopsin from *Halobacterium halobium*.[6] The emission spectrum of intact purple membranes *H. halobium* is characterized by its very short wavelength position (the main maximum is at 314 nm) and can be approximated by two spectral components, one of which corresponds to the azurin spectrum and the other to emission of tryptophan of spectral class I (see below) (Figure 3).

As was mentioned above, spectral form A, the azurin-like spectrum, corresponds to the emission of the unperturbed indole chromophore in a neutral hydrophobic environment. The emission from the ^1L_b state is assumed to contribute more than half of the spectrum.

Spectral form S corresponds to the emission of the indole chromophore located inside the protein globule and forming a 1:1 exciplex with some neighboring polar protein group. Such a spectrum is characteristic, for example, of *Aspergillus* RNAases and L-asparaginase.[3,7] It is of interest that proteins never display the pure class A spectrum; it is always accompanied by a contribution from a class I spectrum (see below), which corresponds to the emission of a 2:1 exciplex (Figure 4). It is assumed that during the excitation lifetime the 2:1 exciplex is formed to a different extent in different proteins depending on the mobility of the chromophore environment. Freezing of the solution down to –196°C does not shift the class S spectra, i.e., the 1:1 exciplex is formed even at low temperatures. The tryptophan fluorescence spectrum of class S possesses a maximum at 316 to 317 nm and shoulders at 305 to 307 and 320 to 330 nm. It is reasonable to think

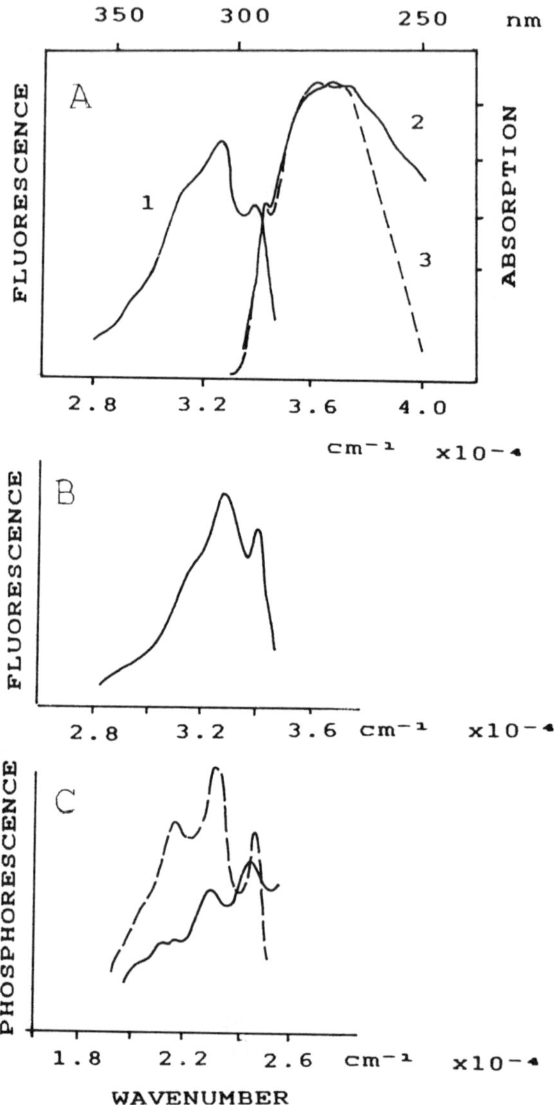

Figure 2. (A) Fluorescence (1), absorption (2), and fluorescence excitation (3) spectra of azurin at 20°C. (B) Fluorescence spectrum of azurin at –196°C. (C) Phosphorescence spectra of azurin (1) and tryptophan in 1 *M* NaCl at –196°C (2). (From Burstein, E. A., Permyakov, E. A., Yashin, V. A., Burkhanov, S. A., and Finazzi-Agro, A., *Biochim. Biophys. Acta*, 491, 155, 1977. With permission.)

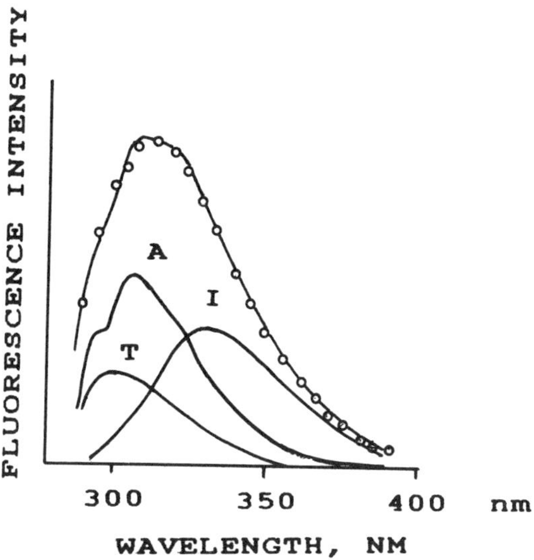

Figure 3. Fitting of experimental spectrum of intact bacteriorhodopsin (points) by theoretical curve, which is a sum of the components corresponding to the emission of tyrosine and tryptophans of spectral classes A and I. (From Permyakov, E. A. and Shnyrov, V. L., *Biophys. Chem.*, 18, 145, 1983. With permission.)

that in all cases when a protein fluorescence spectrum has a shorter wavelength position than that of the class I spectrum, that it contains the S component. It should be noted that such situations are not very common.

Spectral form I corresponds to the emission of indole chromophore located inside the protein globule, forming a 2:1 exciplex with neighboring polar groups of the protein.[4] Tryptophan chromophores of proteins such as actin and chymotrypsin belong to this class. The fluorescence spectrum of class I tryptophans has no vibrational structure; its maximum is located at 330 to 332 nm; its width is 48 to 50 nm (Figure 5).

Spectral form II corresponds to the emission of indole chromophore at the protein surface in contact with bound water molecules, mobility of which in comparison with the mobility of the free water molecules is low. The fluorescence spectrum of class II tryptophan residues is also structureless and has its

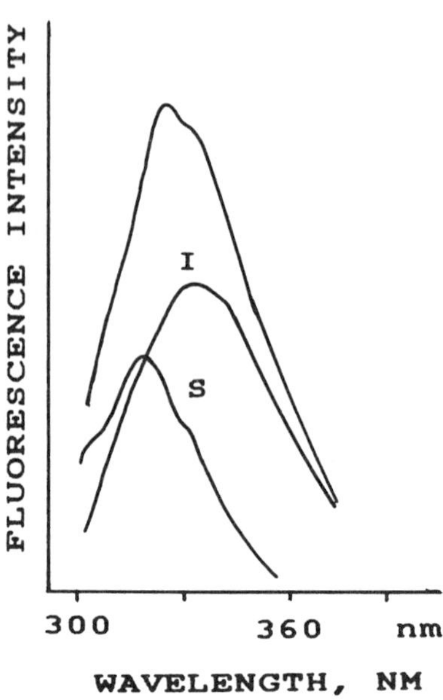

Figure 4. Fluorescence spectrum of L-asparaginase from *E. coli* and its decomposition into S and I components. (From Grischenko, V. M., Emelyanenko, V. I., Ivkova, M. N., Bezborodova, S. I., and Burstein, E. A., *Bioorg. Chem.*, 2, 207, 1976. With permission.)

maximum at 340 to 342 nm and a width of 53 to 55 nm (Figure 5). Many proteins contain tryptophan residues of spectral class II.

Spectral form III corresponds to the emission of indole chromophore located at the protein surface in contact with free water molecules. The spectrum of class III tryptophan residues nearly coincides with the emission spectrum of free tryptophan in water. It is also structureless and has a maximum at 350 to 353 nm; its width is 59 to 61 nm. Tryptophan residues of spectral class III most often occur in unfolded proteins and only sometimes in native ones.

It should be noted that due to their location on the protein surface, the tryptophans of the spectral classes II and III are easily accessible to the solvent and ions and molecules of external quenchers. At the same time, tryptophans of classes A, S, and I are located inside the protein structure and are poorly accessible to water and ions and molecules of quenchers.

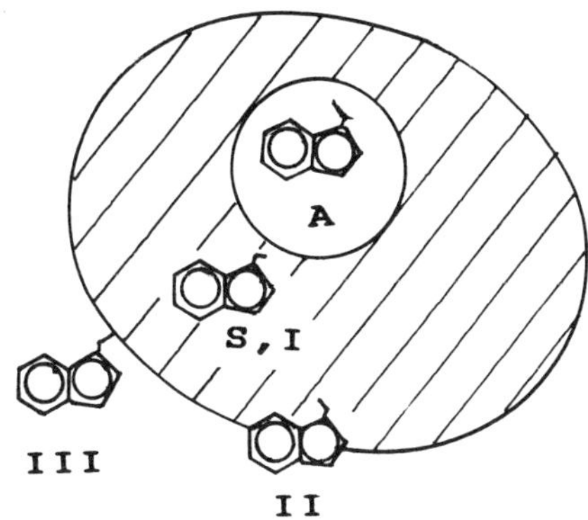

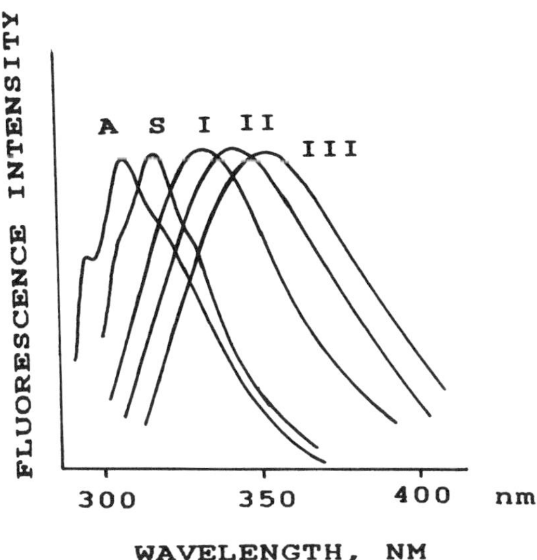

Figure 5. Normalized fluorescence spectra of tryptophans of the spectral classes A, S, I, II, and III. (Upper part) Schematic view of location of tryptophan residues of different spectral classes in a protein globule. (From Burstein, E. A., *Luminescence of Protein Chromophores,* Vol. 6, VINITI, Moscow, 1976. With permission.)

Freezing of the water solution of a protein shifts the fluorescence spectra of tryptophans of classes I, II, and III towards shorter wavelengths due to the immobilization of their polar environment.[8] For class I tryptophans, spectral shifts of 2 to 5 nm

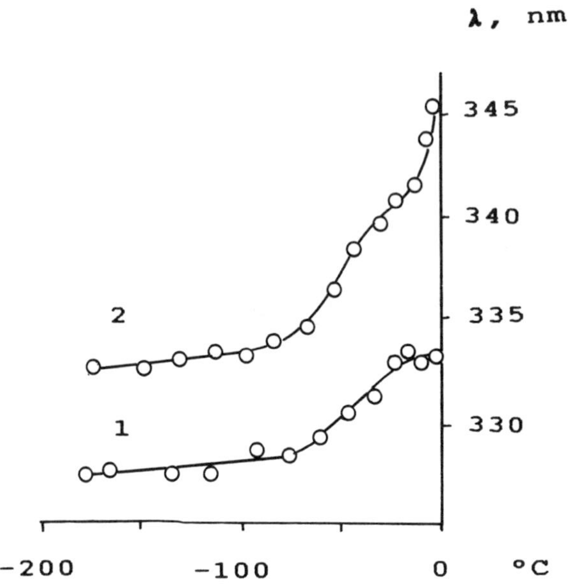

Figure 6. Temperature dependence of fluorescence spectrum position of tryptophans of spectral classes I (1) (β-lactoglobulin) and II (2) (neurotoxin I from cobra venom).

occur within the temperature region from –20 to –90°C (Figure 6). The range over which they occur is about 30°C. For class II and III tryptophans, blue shifts of 5 to 13 nm occur in two stages: the first one within the region from 0 to –20°C, and the second one within the region from –20 to –90°C. The shifts within the region from –20 to –90°C are caused by changes in the relaxational mobility of polar groups of the protein and of bound water molecules, while the shifts at higher temperatures result from changes in the relaxational mobility of weakly bound water molecules.

It should be noted that the model of discrete states of tryptophan residues in proteins is a statistical one, i.e., having tryptophan residues located in environments which result in realization of the A, S, I, II, and III spectral forms is only more probable than having them in other environments. As a matter of fact, there may also be intermediate variants of tryptophan localization, although these are less probable. Nevertheless, analysis of the shape of fluorescence spectra on the basis of the model of discrete states of tryptophan residues in proteins provides useful information as to the location of tryptophan residues in proteins. Such an analysis can be carried out by means of a computer. Using an optimization (fitting) program, experimental spectra of proteins can be fitted by a sum of spectral components corresponding to the emission of tryptophans of spectral classes A, S, I, III, and III by variation of their respective contributions to the total spectrum.

Since the very short wavelength fluorescence spectra which can contain a contribution of A are rarely found, components S, I, II, and III are most often used for such approximation. If the fluorescence is excited by light that also excites tyrosine chromophores, then the tyrosine emission spectrum should be added to the approximating spectral components. Table 1 contains fluorescence spectra of the A, S, I, II, and III components used by us for the analysis of protein fluorescence spectra.

Studies on simulated spectra composed from these components in a known proportion showed that contributions of the components are found by means of the fitting procedure on a computer rather confidently when they lie within the limits from 0.1 to 0.9. Figure 7 shows an example of the fit of the fluorescence spectrum of α-lactalbumin by the S, I, III, and III components.[9] The best fit was achieved by only three components: S, I, and III. The program computes values of the contributions of the components to the total spectrum a_i:

$$a_i = S_i / S \qquad \Sigma a_i = 1 \tag{1}$$

where S_i and S are areas under the spectrum of the ith component and under the total spectrum, respectively. The area under the emission spectrum is proportional to the fluorescence quantum yield value; therefore, one can write

$$a_i = q_i \cdot f_i / q \tag{2}$$

where q_i is fluorescence quantum yield of class i tryptophan, q is mean emission quantum yield of the sample, and f_i is relative concentration of class i tryptophan:

Table 1. Fluorescence Spectra of Tryptophan Residues of Spectral Classes A, S, I, II, and III

λ (nm)	A	S	I	II	III
290	32				
295	68				
300	72	62	14	4	1
305	96	72	31	13	4
310	94	81	47	23	9
315	82	100	66	36	18
320	75	95	82	54	30
325	63	75	95	72	46
330	49	64	100	88	62
335	37	48	99	96	77
340	29	34	90	100	89
345	22	22	81	99	97
350	16	15	70	96	99
355	11	9	59	91	99
360	7	4	49	81	95
365	3	2	40	72	88
370	1	1	33	61	78
375			26	52	70
380			22	43	60
385			17	35	51
390			13	28	43
395			9	24	37
400			7	19	31

$$f_i = n_i / n \qquad (3)$$

n_i is absolute concentration of class i tryptophan; n is total concentration of tryptophans.

Let's present some concrete examples of the analysis of protein fluorescence spectra on the basis of the model of discrete states of tryptophans. Figure 8 shows results of fluorescence study of the process of Ca^{2+} binding to α-lactalbumin from cow milk.[10] The molecule of bovine α-lactalbumin contains four tryptophan residues. The figure demonstrates the dependence of parameters of tryptophan fluorescence of this protein (spectrum position λ and width Δλ and fluorescence quantum yield) on relative concentration of Ca^{2+} and the chelator of divalent cations, EGTA.

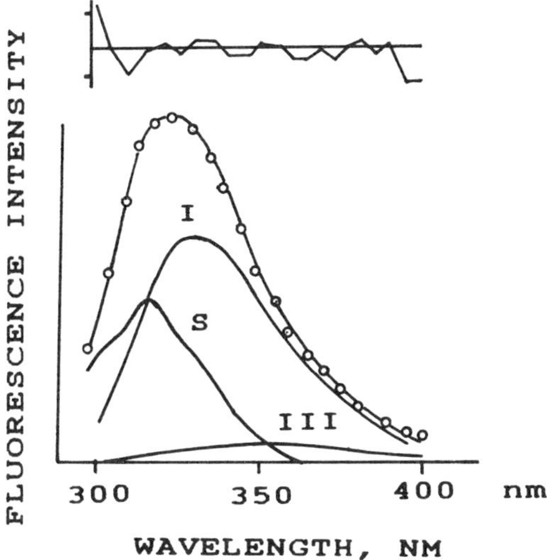

Figure 7. Fitting of experimental fluorescence spectrum of Ca^{2+}-loaded α-lactalbumin by spectral components S, I, II, and III. Points are experimental and the curves are theoretical. The fitting gives only three components. (Upper part) Distribution of residuals.

Figure 7 shows an example of the decomposition of one of the α-lactalbumin fluorescence spectra into components. It is clearly seen in Figure 8A that the binding of Ca^{2+} to α-lactalbumin causes a blue shift of the fluorescence spectrum and its narrowing, which suggests a transfer of some tryptophan residues from the protein surface to a rigid environment in the protein interior. This is well illustrated by the results of the analysis of the spectra on the basis of the model of discrete states of tryptophans in proteins (Figure 8A′): tryptophans of class III (fully accessible to the solvent) become tryptophans of class S (inaccessible to the solvent, forming 1:1 exciplex with a polar group). So the fluorescence method shows that the binding of Ca^{2+} to α-lactalbumin induces conformational changes, resulting in a transfer of some tryptophan residues from the protein surface to internal, solvent-inaccessible regions. It is worth noting that the data of other methods show that this conformational change occurs only on the level of the tertiary structure not concerning its secondary structure.

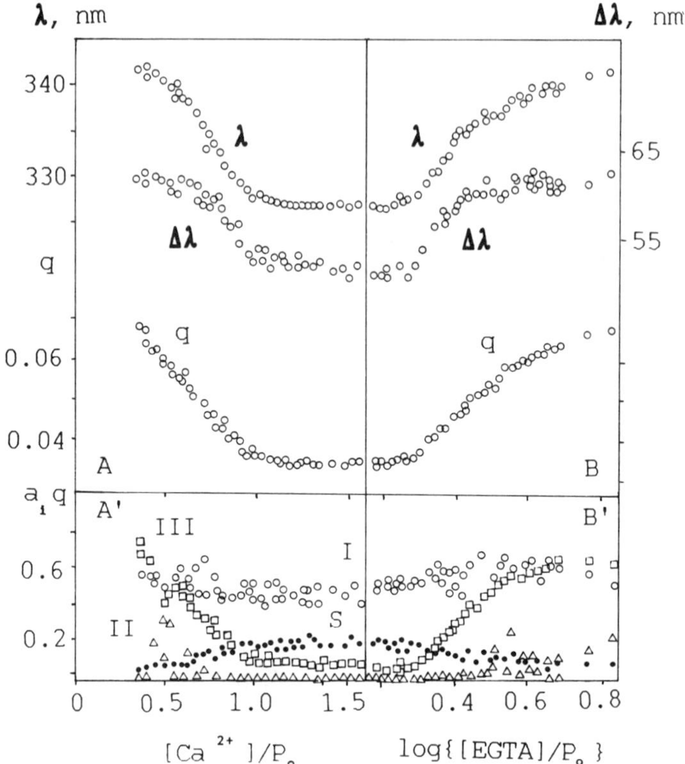

Figure 8. Spectrofluorimetric Ca^{2+} (A) and EGTA (B) titration of bovine α-lactalbumin. λ is fluorescence spectrum position; Δλ is spectrum width; q is fluorescence quantum yield. (A′,B′) the analysis of the fluorescence spectra based on the model of discrete states of tryptophan residues in proteins.[9,10]

Figures 8B and B′ show that the removal of bound Ca^{2+} from α-lactalbumin causes structural changes opposite to those induced by the Ca^{2+} binding.

Figure 9 shows results of a fluorescent study of the thermal denaturation of myosin from bovine skeletal muscle. In this case the thermally induced changes in spectrum position are very small — within the limits of 2 nm — but the shape of the spectrum changes rather fundamentally, which is clearly seen from the considerable changes in the spectrum width. These changes in spectrum shape are clearly seen in an analysis of the spectra on

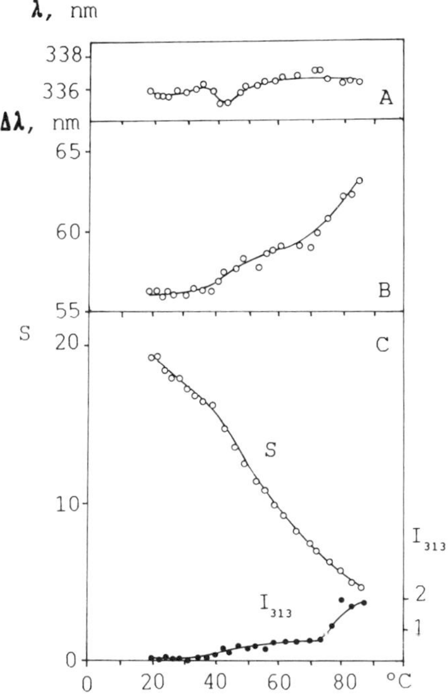

Figure 9. Temperature dependence of fluorescence parameters of myosin from bovine skeletal muscles. (A) Fluorescence spectrum position; (B) spectrum width; (C) relative fluorescence quantum yield. I_{313} is intensity of scattered light of the mercury line at 313 nm.

the basis of the model of discrete states of tryptophans in proteins (Figure 10). The temperature range from 35 to 45°C is characterized by a transition of tryptophans from class I to class S and II (blue spectral shift) accompanied by an increase in light scattering by the protein solution.

When fluorescence is excited by the light of a mercury lamp possessing a line emission spectrum, the increase in light scattering is easy to detect by an increase in the intensity of mercury lines in a scattered light superimposed on the protein emission spectrum (Figure 9C). These changes seem to reflect the process of protein unfolding (appearance of class II tryptophans), complicated by an aggregation of the protein (appearance of class S tryptophans). The next stages of denaturation then proceed:

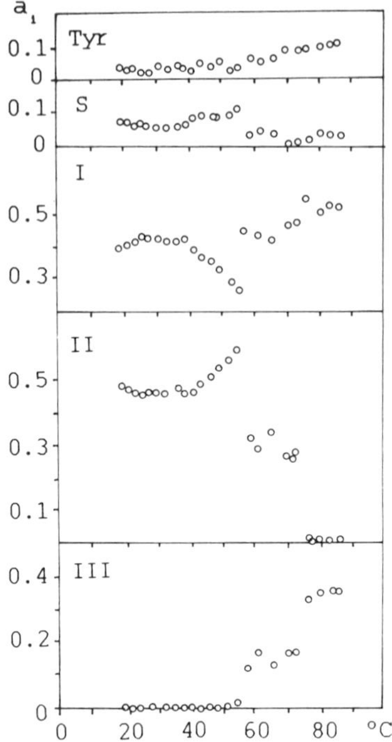

Figure 10. Analysis of myosin fluorescence spectra, corresponding to the temperature dependence presented in Figure 9, based on the model of discrete states of tryptophan residues in proteins.

transition of tryptophans from S and II classes to the I and III classes, the last stage also being complicated by protein aggregation (increase in light scattering and the appearance of class I tryptophans).

So, in this case, in spite of small spectral shifts, the model of discrete states of tryptophans in proteins allows one to analyze the changes of the emission spectrum shape and to obtain valuable qualitative information as to the character of the changes in environment of tryptophan residues in the protein.

As was mentioned above, the fluorescence spectrum of native purple membranes *H. halobium* is approximated by a sum of components corresponding to the emission of tryptophans of the A and I spectral classes (Figure 3). Figure 11 shows the temperature

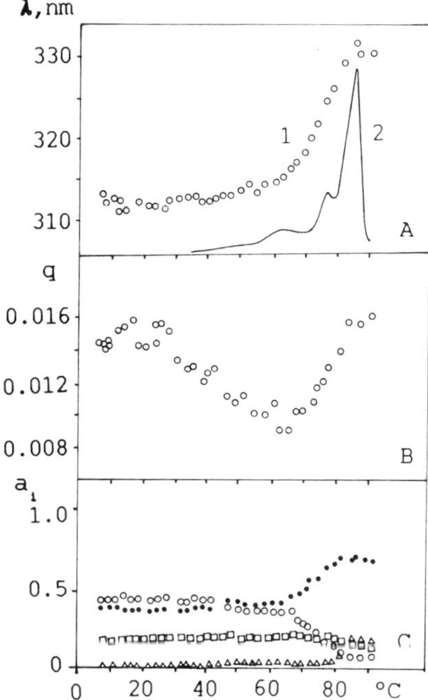

Figure 11. Temperature dependence of fluorescence parameters of bacteriorhodopsin. (A) Fluorescence spectrum position (1); (B) fluorescence quantum yield; (C) contributions of the spectral components of tyrosine (□) and tryptophans of the spectral classes A (○), I (•), and III (Δ). Curve 2 in A is a calorimetric record. (From Permyakov, E. A. and Shnyrov, V. L., *Biophys. Chem.*, 18, 145, 1983. With permission.)

dependence of fluorescence parameters for intact bacteriohodopsin.[6] The red shift of the fluorescence spectrum begins at temperatures above 45°C. There are two stages of the spectral shift: one proceeds at temperatures from 45 to 65°C, and the other one takes place at temperatures from 65 to 90°C. The largest spectral shift occurs within the region from 65 to 90°C and corresponds to the decrease of class A and I tryptophan emissions and the appearance of class III tryptophan emissions, i.e., this temperature region is characterized by a transition of some of the buried tryptophans from a rigid nonpolar environment to a more polar or

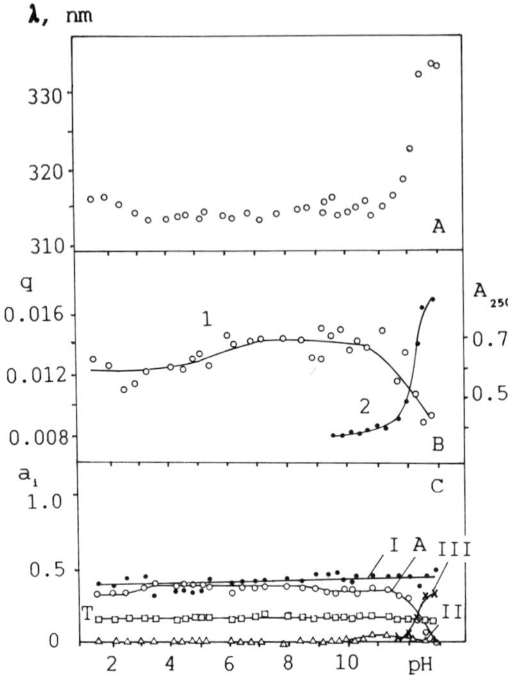

Figure 12. pH dependence of fluorescence parameters of bacteriorhodopsin. (A) Fluorescence spectrum position; (B) fluorescence quantum yield (1) and absorption at 250 nm (tyrosinate) (2); (C) contributions of spectral components T (□), A (○), I (•), II (×), and III (Δ). (From Permyakov, E. A. and Shnyrov, V. L., *Biophys. Chem.*, 18, 145, 1983. With permission.)

even aqueous environment. It is clearly seen that the fluorescence data correlate with the results of microcalorimetry.

Figure 12 shows pH dependence of fluorescence parameters of intact purple membranes measured at 20°C.[6] The red shift of the fluorescence spectrum at pH >11 is caused by alkaline denaturation of bacteriorhodopsin induced by deprotonation of arginines and, at least partially, by titration of phenolic groups of some tyrosine residues with abnormal pK_a values. The appearance of tyrosinate at pH >11 is clearly seen in the absorption spectrum (Figure 12B, curve 2). Analysis of these data in terms of the model of discrete states of tryptophans in proteins shows that the alkali-induced spectral changes correspond to a transfer of some of the

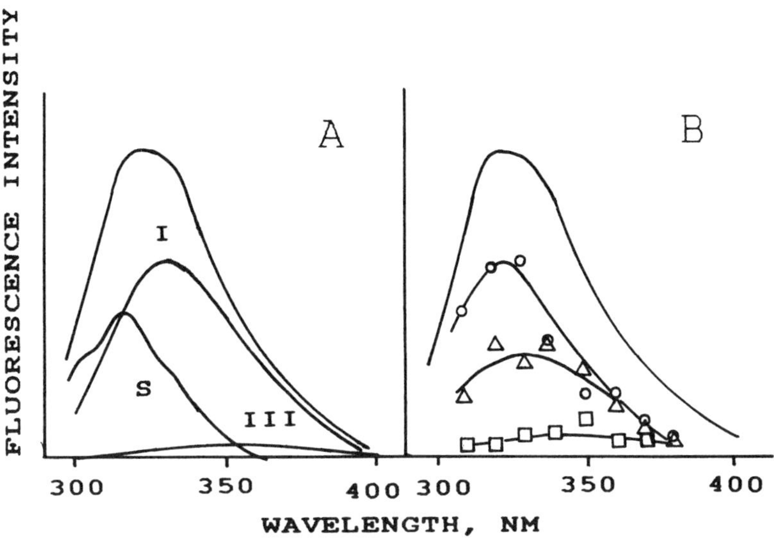

Figure 13. Decomposition of the fluorescence spectrum of bovine α-lactalbumin into components based on the model of discrete states of tryptophans in proteins (A) and on the basis of the time-resolved measurements (B).

buried tryptophan residues to the protein surface in contact with bound water molecules.

In some cases there exists an opportunity to verify the correctness of the results obtained from the spectral analysis on the basis of the model of discrete states of tryptophans in proteins. This can be done by measuring the spectral components by some other method, for example, using the data on fluorescence decay (see below). Figure 13 shows results of the analysis of fluorescence spectra of α-lactalbumin on the basis of the model and on the basis of kinetic measurements (decay-associated spectra). Both methods provide three components (S, I, and III in terms of the model), but with different contributions. Taking into consideration the poor accuracy of determination of the spectral component contributions, one can conclude that the two methods give similar results.

In practice, not every investigation requires the analysis of protein fluorescence spectra in terms of the model of discrete states of tryptophans. Sometimes it is reasonable to decompose

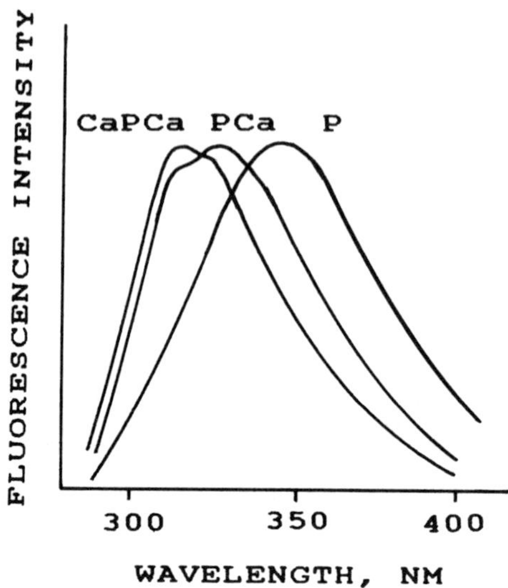

Figure 14. Fluorescence spectra of whiting parvalbumin in the P, PCa, and CaPCa states. (From Permyakov, E. A., Yarmolenko, V. V., Emelyanenko, V. I., Burstein, E. A., Gerday, C., and Closset, J., *Eur. J. Biochem.*, 109, 307, 1980. With permission.)

protein spectra into other components. For example, the binding of two Ca^{2+} ions to whiting parvalbumin occurs according to the scheme

$$P + Ca^{2+} \underset{}{\overset{K_1}{\rightleftharpoons}} PCa$$

$$PCa + Ca^{2+} \underset{}{\overset{K_2}{\rightleftharpoons}} CaPCa \tag{4}$$

where P is the protein. Using fluorescence phase plots (see below) one can obtain the emission spectra of P, PCa, and CaPCa (Figure 14).[11] Each of these spectra is composite and can be analyzed in terms of the model of discrete states of tryptophans, but it is not required for determination of effective Ca^{2+}-binding parameters of the protein.

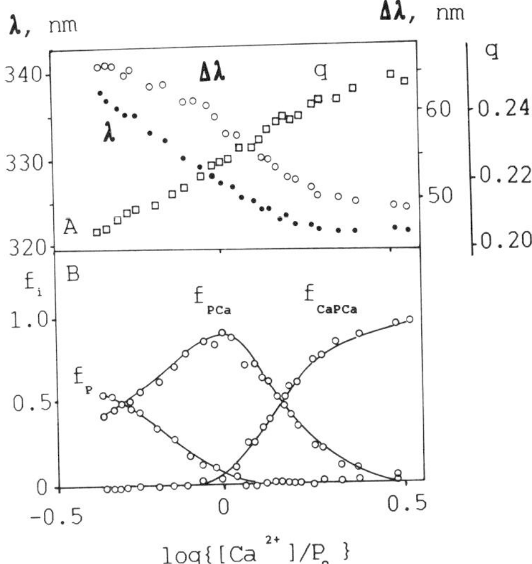

Figure 15. Spectrofluorimetric Ca^{2+} titration of Ca^{2+}-free whiting parvalbumin. (A) Fluorescence spectrum position, λ; spectral width, Δλ; and fluorescence quantum yield q. (B) Populations of the P, PCa, and CaPCa states. Points are experimental and curves are theoretical ones computed according to Scheme 4 and fit to the points by variation of the binding constants. (From Permyakov, E. A., Yarmolenko, V. V., Emelyanenko, V. I., Burstein, E. A., Gerday, C., and Closset, J., *Eur. J. Biochem.*, 109, 307, 1980. With permission.)

Figure 15 shows results of the spectrofluorimetric Ca^{2+} titration of Ca^{2+}-free whiting parvalbumin. The increase in Ca^{2+} concentration causes a blue shift of the fluorescence spectrum, its narrowing, and an increase in fluorescence quantum yield of the protein (Figure 15A). All the spectra were decomposed into components corresponding to the emission of P, PCa, and CaPCa. Figure 15B shows the dependence of the relative populations of these states on Ca^{2+} concentration. The points in the Figure are experimental, while the curves are theoretical ones computed according to Scheme 4 and fitted to the points by variation of K_1 and K_2. The fitting allows evaluation of K_1 and K_2.

There were attempts to work out an independent and objective method of decomposition of composite fluorescence spectra into components corresponding to the genuine emissions of separate

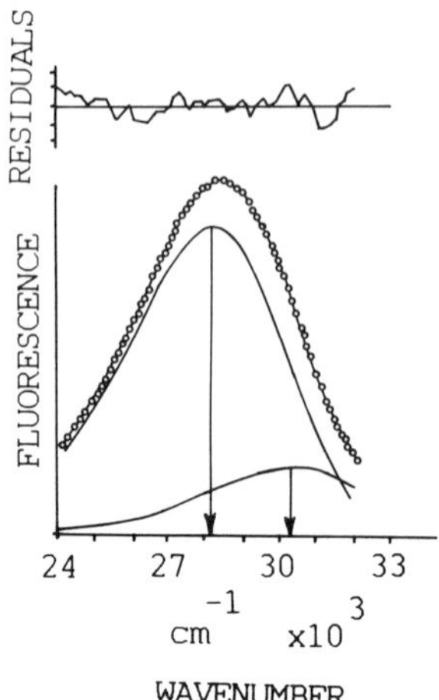

Figure 16. Approximation of fluorescence spectrum of E fragment of fibrinogen by two lognormal components. (Upper panel) Distribution of residuals. (From Emelyanenko, V. I. and Burstein, E. A., in *Proc. 5th Int. Conf. Spectroscopy of Biopolymers,* Kharkov, 1984, 83. With permission.)

tryptophan residues. For example, Emelyanenko et al.[3,12] tried to use the fact that the fluorescence spectra of indole derivatives in various conditions are well described by the lognormal distribution. Though this distribution depends on four parameters (Equation 1, Chapter 3), ν_0, $\nu_{-1/2}$, $\nu_{1/2}$, and I_0, for indole derivatives $\nu_{1/2}$ and $\nu_{-1/2}$ are linearly related to ν_0. For this reason the shape of a smooth, structureless fluorescence spectrum of tryptophan residues in proteins can be determined by a single parameter, ν_0. The fluorescence spectrum of a protein containing several tryptophan residues can be approximated by a sum of several lognormal curves. The fit can be achieved by variation of both the contributions and the positions of the components. Unfortunately, this can only be done for two-component spectra. The method was tested on mixtures of indole derivatives of known content and was shown to provide good results when the components are not too close to each other (up to 5 nm). Figure 16 shows results of the two-component approximation of the fluorescence of the E fragment of fibrinogen (two tryptophan residues per

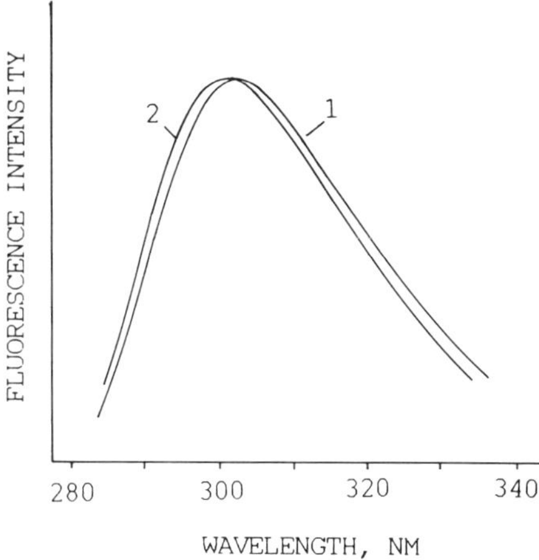

Figure 17. Tyrosine fluorescence spectrum of pike parvalbumin, pI 4.2 (1), and fluorescence spectrum of free tyrosine in water (2).

molecule). For the successful use of this method, the experimental spectrum should be measured with maximum accuracy.

B. **Tyrosine Fluorescence**

Proteins devoid of tryptophan but possessing tyrosine residues occur rather often. Calmodulin, troponin C, some parvalbumins, and ribonuclease A belong to this class of proteins. Their fluorescence spectra differ almost not at all from the spectrum of free tyrosine in water (maximum at 303 to 306 nm). Denaturation of the proteins by urea, acidic or alkaline pH, or temperature produce practically no change in spectrum shape and position. Figure 17 shows fluorescence spectrum of pike parvalbumin pI 4.2 containing one tyrosine residue per molecule. For comparison, the same figure also demonstrates the spectrum of an aqueous solution of tyrosine.

The freezing of the protein solution shifts the tyrosine spectrum toward shorter wavelengths by 3 to 4 nm, and the shift can proceed in two stages, just as in the case of tryptophan fluorescence (Figure 18). This seems to be caused by the freezing of the fast nanosecond mobility of different parts of the protein-bound water system.

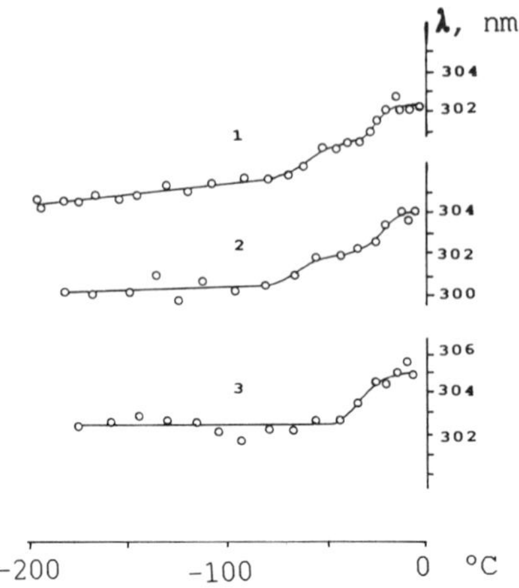

Figure 18. Temperature dependence of the tyrosine fluorescence spectrum position for carp parvalbumin, pI 3.95 (1), rabbit skeletal muscle troponin C (2), and cobra venom cytotoxin (3).

Fluorescence spectra of proteins containing both tyrosine and tryptophan residues usually result mainly from the emission of indole chromophores. Nevertheless, some of these proteins display a rather clear tyrosine component. Figure 19 shows fluorescence spectrum of troponin I from bovine skeletal muscle which contains an essential tyrosine component. To isolate the tyrosine component, one can measure the protein fluorescence spectra at two different excitations: for example, at 280 nm when both tyrosine and tryptophan absorb and at 297 nm, when the absorption is due only to tryptophan residues. The difference between these spectra will be the tyrosine contribution.

C. **Phenylalanine Fluorescence**

Proteins containing phenylalanines, but devoid of both tyrosine and tryptophan residues, occur rather seldom. Most parvalbumins and green pea superoxide dismutase belong to this group of proteins. Fluorescence spectra of phenylalanine residues in proteins display a distinct vibrational structure just like that of aqueous phenylalanine solution (Figure 20), but the

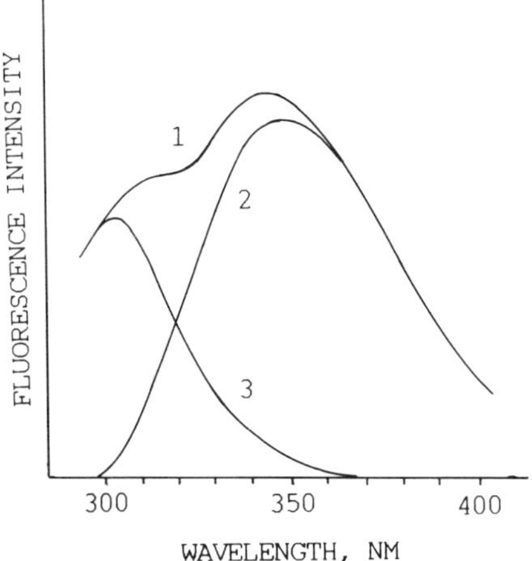

Figure 19. Fluorescence spectra of troponin I from bovine cardiac muscles measured at excitation 280.4 nm (1) and 296 nm (2). (3) The difference between the spectra 1 and 2.

maxima in the protein spectrum are 2.0 to 2.5 nm red shifted. A similar shift is observed in the absorption spectrum too. These differences are thought to be caused not by the change in the solvent polarity and specific interactions, but by a change of the inductive effect of the substituent side chain of the benzene ring. In spite of the very high phenylalanine content in these proteins (for example, parvalbumins having molecular mass 12,000 contain 10 to 11 phenylalanine residues per molecule) and the rather small distances between the chromophores, we did not find any traces of the excimer fluorescence, which is characteristic for polyphenylalanine, for example.

The freezing of the protein solution does not cause any shifts of phenylalanine fluorescence spectra, but makes the vibrational structure clearer. The resolution of the phenylalanine spectrum of hake parvalbumin (parameter $R = I_{283}/I_{288} - 1$) changes over a wide temperature region from –30 to –130°C (Figure 21). The curve reflects the change in populations of vibronic levels of the chromophore.

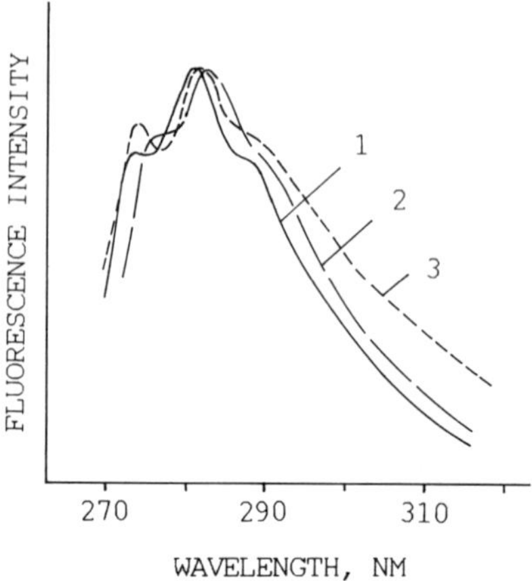

Figure 20. Fluorescence spectra of free phenylalanine in water (1) and phenylalanine residues in carp parvalbumin, pI 4.47 (2), and green pea superoxide dismutase (3) at 20°C.

III. FLUORESCENCE QUANTUM YIELDS OF PROTEINS

A. Trytophan Fluorescence

The values of quantum yields of tryptophan fluorescence of proteins vary over wide limits. They may be both higher and lower than the fluorescence quantum yield for free tryptophan in water (0.23 to 20°C[1]) . There is no good correlation between protein spectrum position and fluorescence quantum yield.

Changes in location of tryptophan residues during the course of conformational changes can both increase and decrease fluorescence yield. For example, the binding of Ca^{2+} to bovine α-lactal-bumin causes changes in its structure such that some of its tryptophan residues are transferred from the protein surface to the interior of the protein globule in a rigid hydrophobic environment (Figure 22A). This is accompanied by a decrease in fluorescence quantum yield, since the tryptohan residues enter an environment containing effective quenching groups, which may be the –S–S– bridges. A similar change in location of the tryptophan residue in whiting or cod parvalbumin (transfer from the surface to the interior) results in a twofold

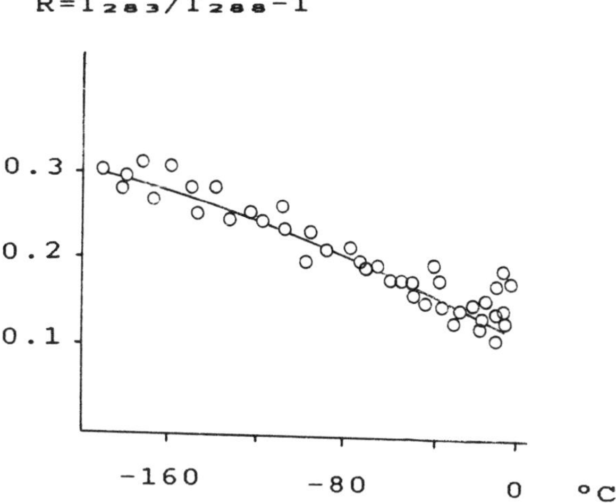

Figure 21. Temperature dependence of the resolution of the vibrational structure in phenylalanine fluorescence spectrum of hake parvalbumin.

increase in fluorescence quantum yield (Figure 22B). In this case the environment of trytophan does not contain highly effective groups. On the contrary, this transition is accompanied by elimination of the fluorescence quenching by water.

B. **Tyrosine Fluorescence**

As a rule, the quantum yield of tyrosine fluorescence of proteins is lower than the fluorescence yield of free tyrosine in water (0.21 at 20°C[1]). The peptide is thought to be a universal quencher of tyrosine fluorescence in proteins. Sometimes the fluoresence of some tyrosine in proteins may be almost completely quenched.

One can think that the surface tyrosine residues in proteins display unquenched fluorescence, the quantum yield of which is equal to the fluorescence quantum yield of free tyrosine in water. The quenching of tyrosine fluorescence occurs due to its interactions with the hydrated carboxyls of peptide groups (dynamic quenching), with ionized tyrosine in alkaline medium (excitation energy transfer), with disulfide groups (contact quenching), with peptide carboxyls or carbonyls (formation of hydrogen bond), and with carboxylates (collisional quenching).

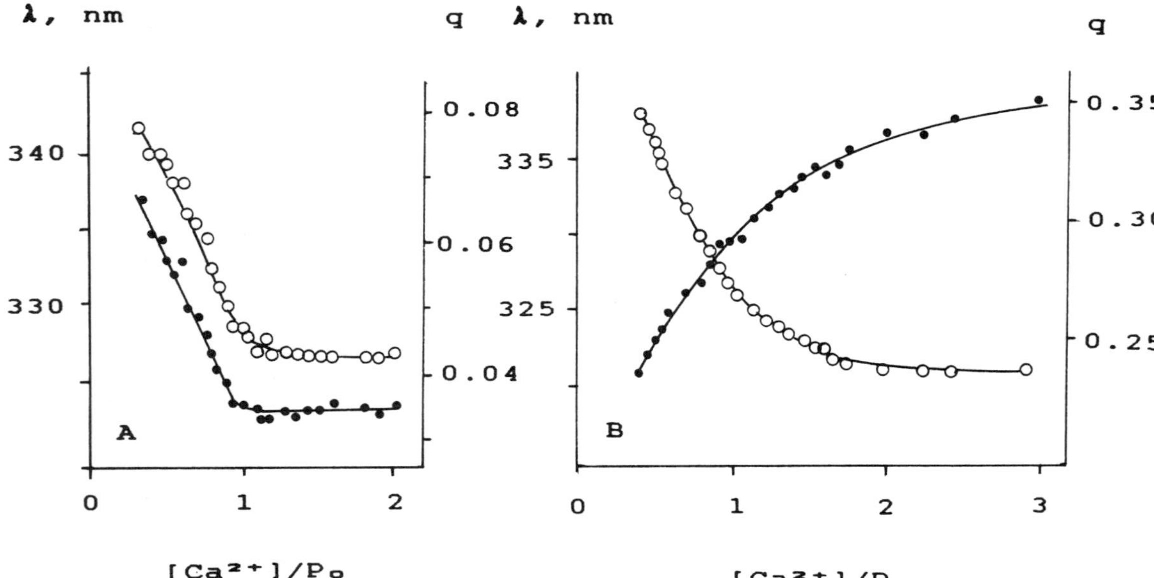

Figure 22. The opposite changes of tryptophan fluorescence quantum yield caused by similar changes in the location of tryptophan residues induced by Ca^{2+} binding to bovine α-lactalbumin (A) and whiting parvalbumin (B).

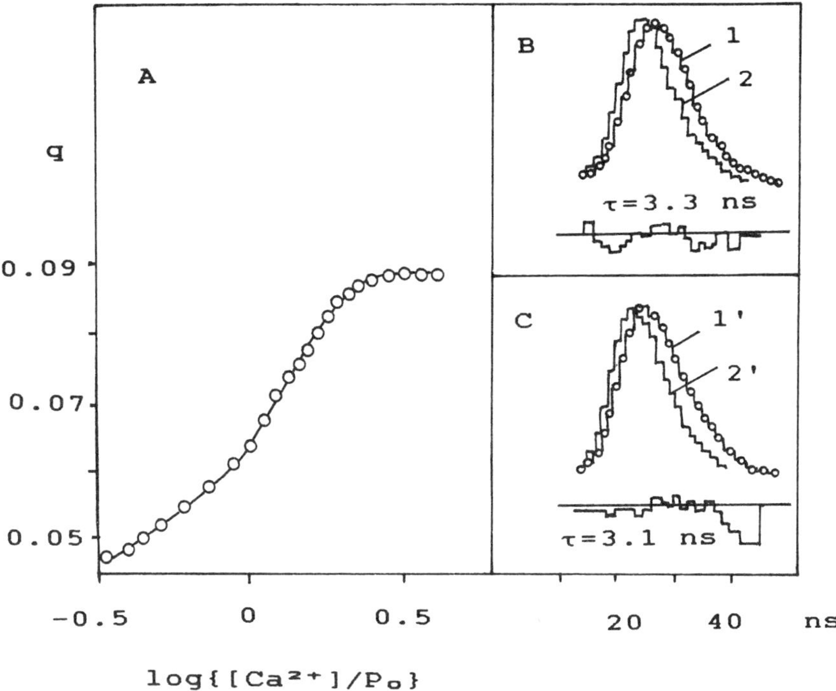

Figure 23. (A) Spectrofluorimetric Ca^{2+} titration of carp parvalbumin, pI 3.95. (1,1′) fluorescence decay curve for Ca^{2+}-loaded (B) and Ca^{2+}-free (C) carp parvalbumin, pI 3.95. (2,2′) profiles of excitation impulses obtained from synchrotron. (Below) Distribution of residuals. (From Permyakov, E. A., Burstein, E. A., Emelyanenko, V. I., Alexandrov, Y. M., Glagolev, K. V., Makhov, V. N., Syreishchikova, T. N., and Yakimenko, M. N., *Biofizika,* 28, 393, 1983. With permission.)

Often, changes in the tyrosine fluorescence yield of proteins proceed without changes in fluorescence lifetime. For example, in the case of carp parvalbumin, pI 3.95, containing one tyrosine residue per molecule, the binding of Ca^{2+} causes a twofold increase in fluorescence yield, but does not change the fluorescence lifetime (Figure 23).[13] The fluorescence quenching in this case seems to be caused by formation of a nonfluorescent complex of the tyrosine residue with a neighboring quenching group. The binding of Ca^{2+} to the protein changes its structure and shifts the equilibrium in favor of the free chromophore, the fluorescence of which we detect.

C. PHENYLALANINE FLUORESCENCE

The quantum yield of protein phenylalanine fluorescence can be both higher and lower than the fluorescence yield of free phenylalanine in water (0.038 at 20°C[1]). For example, the quantum yields of phenylalanine fluorescence of carp, pike, and hake parvalbumins are higher than the emission yield of free phenylalanine, and for carp parvalbumin, pI 4.37, the fluorescence yield equals 0.27. The pronounced increase in fluorescence yield upon the inclusion of phenylalanine into the protein seems to be caused by elimination of fluorescence quenching by water. At the same time, the quantum yield of the phenylalanine fluorescence of superoxide dismutase containing Cu^{2+} and Zn^{2+} ions is about 0.01 and increases up to 0.03 to 0.05 only after acidic or alkaline denaturation of the protein.

IV. FLUORESCENCE DECAY OF PROTEINS

Since fluorescence decay times of proteins lie in the nanosecond region, their measurement is a rather difficult problem. These measurements can be carried out, for example, using pulsed excitation light sources and multichannel analyzers of pulses for statistical detection of weak light signals.[14,15] The observed kinetics of fluorescence decay, F(t), is a convolution of instrument response function, A(t), and genuine decay function, f(t) (which is emission decay upon excitation by δ-function):

$$F = \int A(t) f(t - x)\,dx \tag{5}$$

A scatterer is used for determination of the response function. There is no general analytical method of determination of f(t) from F(t) and A(t). Usually f(t) is assumed to be a sum of exponentials

$$f(t) = \Sigma a_i \cdot \exp(-t / \tau_i) \tag{6}$$

where a_i and τ_i are contribution and lifetime of i component, respectively.

Several methods of deconvolution have been proposed such as, for example, use of Laplace or Fourier transforms, use of the method of moments, and so on. The most widespread method of deconvolution is the least-squares method in which the experimental decay kinetics is fit by a theoretical curve (Equation 5) with variation of a_i and τ_i. The criterion of quality of the fit is the X^2 criterion and the randomness of the distribution of differences between experimental and theoretical values.

It is evident that by using this method one can decompose an experimental decay curve into two or, at most, three exponents; therefore, it is reasonable to measure the fluorescence decay curve only for proteins containing a small number of chromophores.

A. Tryptophan Fluorescence

Tryptophan fluorescence lifetimes are usually within the nanosecond region. For proteins with two to three tryptophan residues, it is possible sometimes to detect several exponentials in the decay curve and to identify them with specific chromophores. The situation can be complicated, however, by the existence of protein conformers. For example, the fluorescence decay curve for the single tryptophan residue in whiting parvalbumin (Figure 24) is adequately approximated only by a sum of two exponentials (Figure 25). This fact can be interpreted either as a reflection of an essential nonexponentiality of the fluorescence decay of the single tryptophan in this protein or as the existence of two real components caused by emission of two protein conformers. The fact is that the steady-state fluorescence spectra of the protein at any saturation by Ca^{2+} are wider than would be expected for the single tryptophan residue, i.e., the spectra seem to contain at least one more component. Measuring fluorescence decay at different wavelengths and having the steady-state fluorescence spectrum of the protein $F(\lambda)$, one can obtain emission spectra of the conformers $F_1(\lambda)$ and $F_2(\lambda)$ (decay-associated spectra):

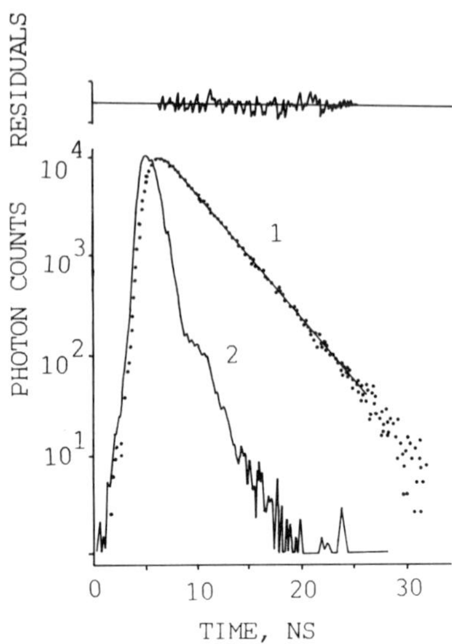

Figure 24. Fluorescence decay curve for whiting parvalbumin (1). Points are experimental and the curve is a double-exponential approximation.[16] (2) Excitation impulse. (Upper panel) Distribution of residuals.

$$F(\lambda) = F_1(\lambda) + F_2(\lambda) \tag{7}$$

$$F_1(\lambda) = \frac{A_1(\lambda) \cdot \tau_1(\lambda) \cdot F(\lambda)}{A_1(\lambda) \cdot \tau_1(\lambda) + A_2(\lambda) \cdot \tau_2(\lambda)} \tag{8}$$

$$F_2(\lambda) = \frac{A_2(\lambda) \cdot \tau_2(\lambda) \cdot F(\lambda)}{A_1(\lambda) \cdot \tau_1(\lambda) + A_2(\lambda) \cdot \tau_2(\lambda)} \tag{9}$$

where A_1, A_2, and τ_1, τ_2 are contributions and lifetimes of the components 1 and 2, respectively. Results of such an analysis for Ca^{2+}-loaded whiting parvalbumin are presented in Figure 26. It is clearly seen that the total emission spectrum consists of two components differing in their maximum position. The blue

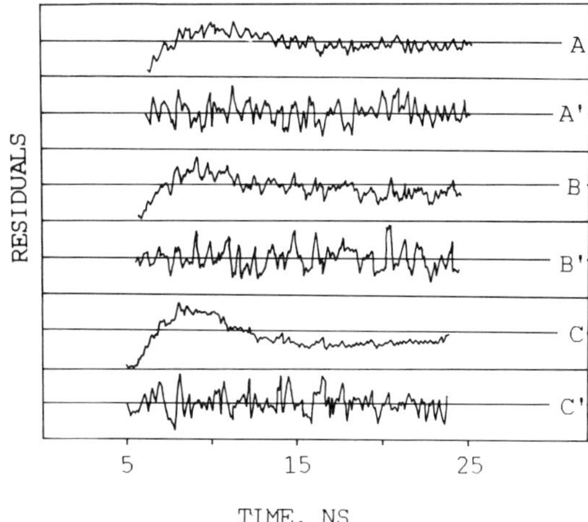

Figure 25. Distribution of residuals for the fitting of the fluorescence decay for whiting parvalbumin by one (A,B,C) and two (A′,B′,C′) exponentials.[16] (A,A′) The protein with two bound Ca^{2+} ions; (B,B′) the protein with one bound Ca^{2+} ion; (C,C′) apoprotein.

component seems to correspond to a more compact conformer, while the red one corresponds to a looser, more relaxed conformer. Such an interpretation is corroborated by the fact that ions of the external quencher I^- only quench the fluorescence of the relaxed conformer, and that the substitution of H_2O by D_2O shifts the equilibrium between conformers in favor of the more compact one.

The conclusion as to the existence of conformers was obtained also for one more protein with a single tryptophan residue, azurin.[17]

Bovine α-lactalbumin contains four tryptophan residues per molecule. Due to their vicinity to disulfide bonds, their fluorescence of some of them is quenched. Figure 27 shows the fluorescence decay curve of Ca^{2+}-loaded α-lactalbumin. The decay data cannot be approximated by a single exponential. The result of a double-exponential approximation depends on the region of the fit: there appears to be at long times a long-living component. For this reason we were forced to approximate the decay data by a

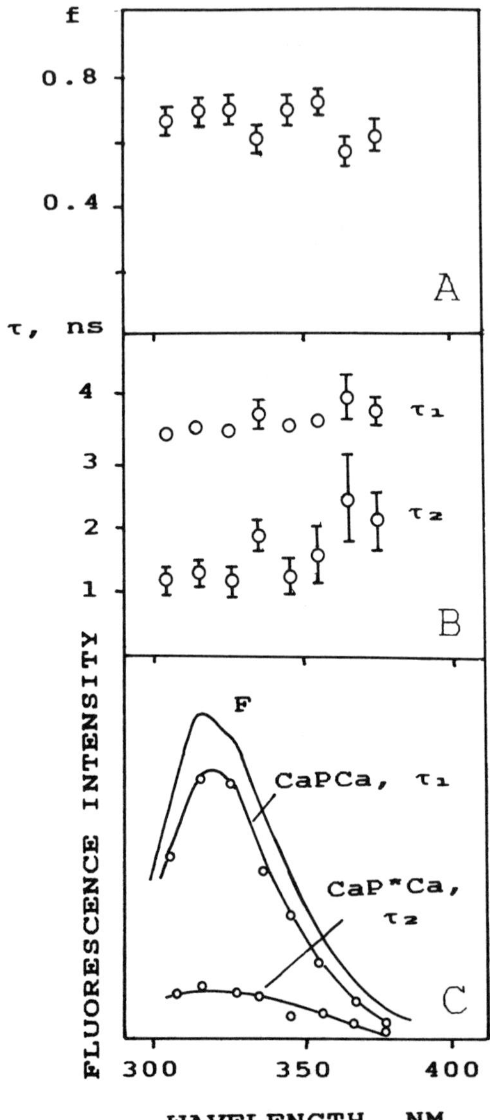

Figure 26. Wavelength dependence of relative contribution of the emission of one of two components (A) and of lifetimes τ_1 and τ_2 of the components (B) for whiting parvalbumin. (C) Total emission spectrum and components corresponding to emission of two parvalbumin conformers. (From Permyakov, E. A., Ostrovsky, A. V., Burstein, E. A., Pleshanov, P. G., and Gerday, C., *Arch. Biochem. Biophys.*, 240, 781, 1985. With permission.)

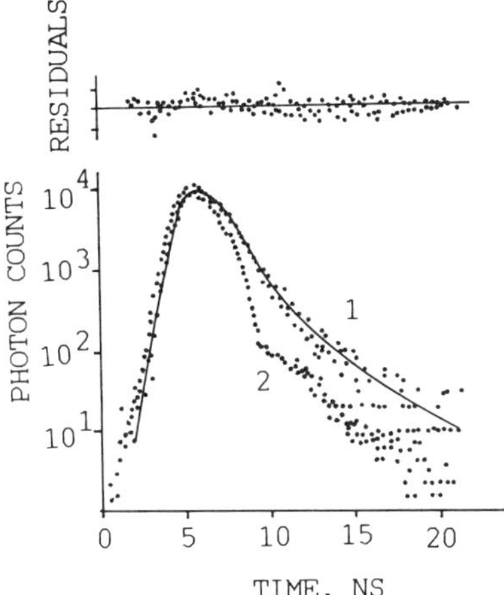

Figure 27. (1) Fluorescence decay for α-lactalbumin. Points are experimental, the curve is theoretical (approximation by three exponents). (2) Excitation impulse. (Upper part) Distribution of residuals.

sum of three exponential components. Figure 28A and B shows the wavelength dependence of the contributions and lifetimes of the components. Figure 28C shows the emission spectra of the components. In terms of the model of discrete states of tryptophan residues in proteins, these roughly correspond to the spectra of the S, I, and III forms. As was mentioned above, the analysis of the spectrum on the basis of the model of discrete states of tryptophans gives the same components, but with other contributions.

B. **Tyrosine Fluorescence**

Values of lifetimes for tyrosine fluorescence of proteins usually lie in the nanosecond range. Since the position of the tyrosine fluorescence spectrum is practically independent of the chromophore environment, it is hard to find contributions of individual tyrosine residues to the total spectrum. Nevertheless, measurements of tyrosine fluorescence decay can also provide useful information. For example, the study of the emission decay of the single tyrosine residue in pike parvalbumin, pI 4.2, which

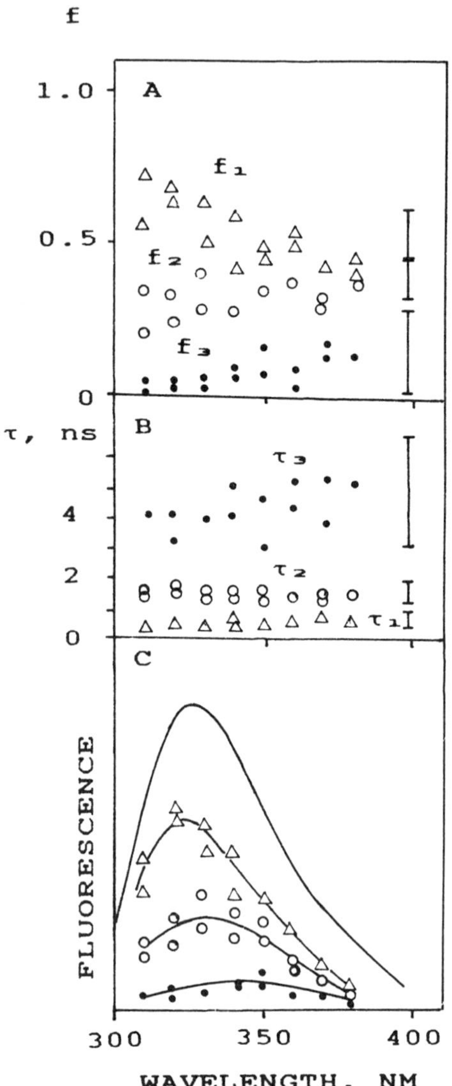

Figure 28. Wavelength dependence of contributions (A) and lifetimes (B) of three exponential components approximating fluorescence decay of α-lactalbumin. (C) Emission spectra of the components.

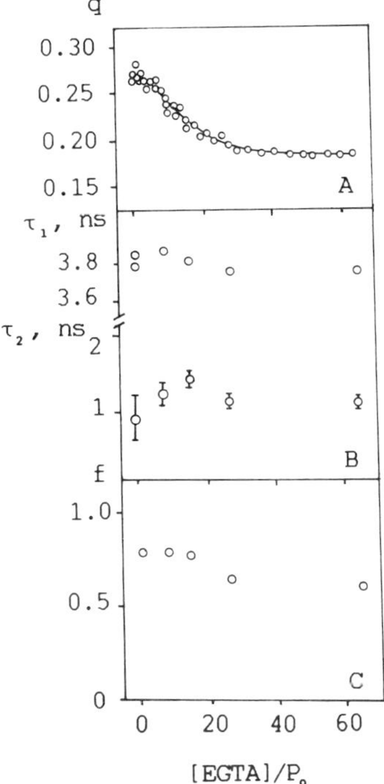

Figure 29. Spectrofluorimetric EGTA titration of pike parvalbumin, pI 4.2. (A) Fluorescence quantum yield; (B) fluorescence lifetimes for two emission components; (C) contribution of the first component. Fluorescence was excited at 280.4 nm.

is fit by two exponential components, allowed one to draw a conclusion as to the existence of two conformers of the protein at any Ca^{2+} concentration in solution. The change in Ca^{2+} concentration alters the lifetime and contributions of the emission of the conformers (Figure 29). Excitation of the fluorescence of the tyrosine residue in pike parvalbumin, pI 4.2, at 257 nm, where phenylalanine residues give the main contribution to the absorption, increases τ_1 and τ_2 and makes the contribution of component 2 negative (Figure 30). This suggests the existence of excitation energy transfer from phenylalanines to the tyrosine. The increase in the lifetimes of the components and the negative contribution of one of them are explained by a rather large lifetime

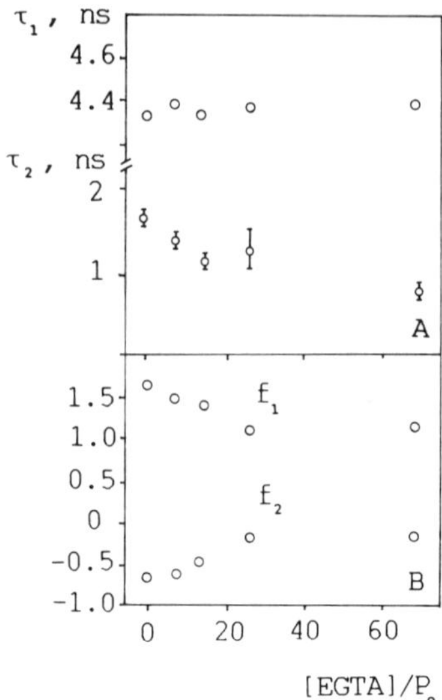

Figure 30. Spectrofluorimetric EGTA titration of pike parvalbumin, pI 4.2. (A) Fluorescence lifetimes for two emission components; (B) contributions of the components to the total emission. Fluorescence was excited at 257 nm.

of the phenylalanine excited state. Thus, the measurements allow detection of the presence of excitation energy transfer from one type of chromophore to another, which can be of importance for the interpretation of steady-state fluorescence data and for evaluation of distances between chromophores.

C. Phenylalanine Fluorescence

Lifetimes of phenylalanine fluorescence of proteins are within the region from several nanoseconds to several tens of nanoseconds. Fluorescence decay for phenylalanine in proteins has not been studied very much, which is explained by the rarity of proteins devoid of tyrosine and tryptophan residues. An example of such measurements is the measurement of the fluorescence decay of phenylalanine residues in carp parvalbumin, pI 4.47 (Figure 31).[13] The decay curves for Ca^{2+}-loaded parvalbumin are approximated by a sum of two exponentials. τ_1 and τ_2 obtained from the fit differ by more than an order of magnitude (5.4 and

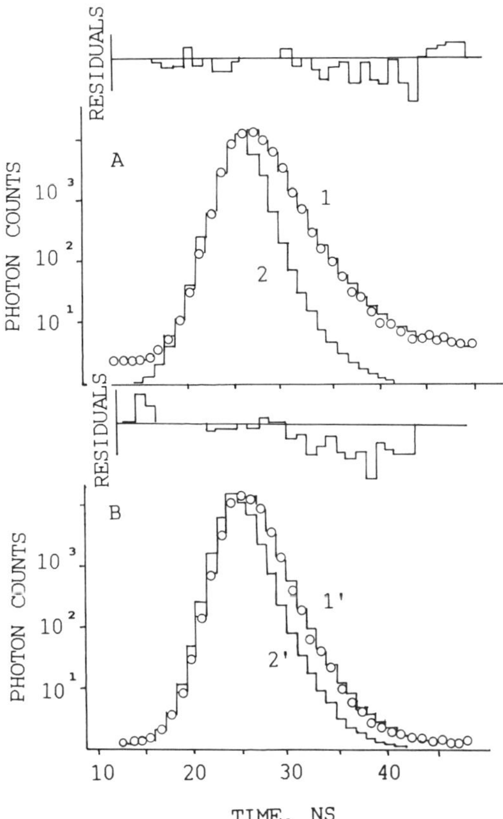

Figure 31. Fluorescence decay for phenylalanine residues in carp parvalbumin. (1,1′) Decay curves; (2,2′ excitation impulses obtained from synchrotron. (Upper parts) Distribution of residuals. (A) Ca^{2+}-loaded protein (double-exponential approximation); (B) apoprotein (single-exponential approximation). (From Permyakov, E. A., Burstein, E. A., et al., *Biofizika*, 28, 393, 1983. With permission.)

62 ns). According to the X-ray data, ten phenylalanines in carp parvalbumin are located at rather short distances from each other (<7.5 Å). The distance of 50% resonance excitation energy transfer for the donor-acceptor pair Phe-Phe in aqueous solution is 5.5 Å and is 1.1 to 1.4 times longer in proteins. This would suggest that an effective excitation energy transfer occurs between the Phe residues in parvalbumin. The presence of the two exponentials in

the decay curve seems to show that the system of ten Phe residue benzene rings can be separated into two subsystems emitting almost independently of each other. Deactivation processes seem to be absent in one of them. It may be that this system comprises phenylalanine residues of the hydrophobic core of the molecule. Ca^{2+} release disturbs this subsystem.

V. GENERAL STRATEGY OF LUMINESCENT STUDY OF PROTEINS

The intrinsic fluorescence method can be used for various purposes. First, it can provide valuable structural information about proteins, though this information is local since it reflects the structure and properties of the immediate environment of protein chromophores. Second, using changes in protein fluorescence induced by various effectors one can draw conclusions as to the character of structural changes caused by the effectors. Third, some fluorescence parameters can be used for determination of various physical and physicochemical characteristics of proteins. In the first two cases, the number of chromophores in a protein should not be too large in order to facilitate interpretation of the experimental results.

It should be noted that intrinsic protein fluorescence can be far from sensitive to every event in a protein. There may be situations when the aromatic amino acid residues which provide luminescent information are located in regions wherein the structure and mobility of which are not affected by events in other parts of the protein.

In starting to study a new protein by the luminescence method, one should elucidate its stability to the action of such universal factors as temperature, pH, and ionic strength and find possible temperature-, pH-, and ionic strength-induced structural changes. In the absence of such information we could find ourselves in a situation in which all further measurements would be carried out at values of the parameters indicated such that the protein is passing from one state to another, i.e., on a mixture of different protein states. This might complicate and entangle interpretation of the experimental data obtained.

After obtaining information as to the effects of temperature, pH, and ionic strength on fluorescence parameters of the protein, one can start to study effects of various specific factors on protein structure. They can be various metal ions, low- and high-molecular mass, biologically active compounds, other proteins, and so on. The choice of the factors depends on available information on the functions of the given protein. If the protein fluorescence is sensitive to these factors, one can try to measure the parameters of their interactions with the protein.

Sometimes, in order to obtain information on the protein structure, it is useful to study its stability against the action of various denaturants, such as, for instance, urea or guanidine hydrochloride, and of solvents such as alcohols or dioxane.

It should be noted that in order to get quantitative information on parameters of conformational changes in a protein or to measure parameters of its interactions with other compounds, one should use fluorescence parameters which are linearly related to the fraction of conversion from one state to another. Moreover, one should be able to detect possible intermediate states of the protein in the course of its transition from one state to other. All these methods are presented elsewhere in this book.

VI. FLUORESCENCE PARAMETERS AS A MEASURE OF THE FRACTION OF CONVERSION FROM ONE STATE TO ANOTHER IN PROTEINS

In order to use a parameter of protein fluorescence for quantitative measurements, one must know its dependence on the fraction of conversion in the protein.[2,3] If the protein passes from a P state to a P_1 state ($P \rightarrow P_1$), the fraction of the P to P_1 conversion, α, is

$$\alpha = [P_1] / ([P] + [P_1]) \tag{10}$$

where [P] and $[P_1]$ are concentrations of P and P_1, respectively.

Spectrum position or spectrum width cannot serve as a measure of fraction of conversion, since their dependence on α can be rather complex and poorly determinable. When the initial and final states of a protein differ significantly and their contributions to the total spectrum are comparable, the spectrum width at intermediate α is, as a rule, larger than that of the initial and final spectra. In many cases, changes of spectrum maximum position are also not proportional to α. This is especially visible when emission quantum yields of the initial and final states essentially differ. Strictly speaking, the fluorescence quantum yield is also nonlinearly dependent on the fraction of conversion because of possible changes in absorption of the excitation light during the course of the $P \rightarrow P_1$ transition. Nevertheless, in most cases the changes in protein absorption are negligibly small and can be taken into account (see below). The changes in fluorescence yield can still be used as a measure of fraction of conversion:

$$q = q_p + \alpha \cdot (q_{pl} - q) \qquad (11)$$

Since fluorescence intensity at a fixed wavelength I_λ is linearly related to q, it can also be used as a measure of fraction of conversion:

$$I_\lambda = I_{\lambda,p} + \alpha \cdot (I_{\lambda,pl} - I_{\lambda,p}) \qquad (12)$$

In contrast to fluorescence intensity at a fixed wavelength, intensity at the wavelength of the spectrum maximum cannot be used as a measure of α. This is connected with the fact that this parameter does not take into account possible changes in the shape of the spectrum, and when such changes take place they do not occur proportionally to q and α.

VII. SEARCH FOR INTERMEDIATE STATES

A convenient method for searching for intermediate states of a protein during the process of its transition from initial to final state is analysis of the so-called fluorescent phase plots.[4] In order to draw such a plot one should use two relatively independent parameters linearly related to α. Such parameters can be, for example, fluorescence intensities at two different wavelengths $I_{\lambda 1}$ and $I_{\lambda 2}$:

$$I_{\lambda 1} = A_1 + B_1 \cdot \alpha; \qquad I_{\lambda 2} = A_2 + B_2 \cdot \alpha \tag{13}$$

If the transition proceeds between two states, then $I_{\lambda 1}$ and $I_{\lambda 2}$ are linearly related to each other:

$$I_{\lambda 1} = (A_1 - B_1 / B_2 \cdot A_2) + B_1 / B_2 \cdot I_{\lambda 2} \tag{14}$$

It is easy to demonstrate that in the case of the existence of an intermediate or parallel product that possesses $I_{\lambda 1}$ and $I_{\lambda 2}$ which are different from those for the initial and final states, the relationship between $I_{\lambda 1}$ and $I_{\lambda 2}$ becomes nonlinear.

Consider some examples of the use of the fluorescent phase plots. Figure 32B shows a fluorescent phase plot corresponding to the EGTA titration of Ca^{2+}-loaded whiting parvalbumin. It is clearly seen that the transition does not occur between only two states. A break on the curve corresponds to the maximal population of an intermediate state. Parvalbumins bind two Ca^{2+} ions per molecule; therefore, the intermediate state revealed by the fluorescent phase plot corresponds to the protein with one bound Ca^{2+} per molecule — PCa. Figure 32C shows emission spectra of the initial CaPCa, intermediate PCa, and final P states of the protein. Figure 32A demonstrates that the EGTA-induced changes in spectrum position and fluorescence yield are smooth, and the intermediate state can be revealed only by the fluorescent phase plots.

Figure 33B shows a fluorescent phase plot corresponding to the urea-induced denaturation of Ca^{2+}-loaded α-lactalbumin. The break on the plot corresponds to the maximal concentration of an intermediate state of α-lactalbumin. It should be noted that the existence of the intermediate state is clearly seen also only in the fluorescent phase plot and not in the simple dependencies of λ and q on urea concentration (Figure 33A).

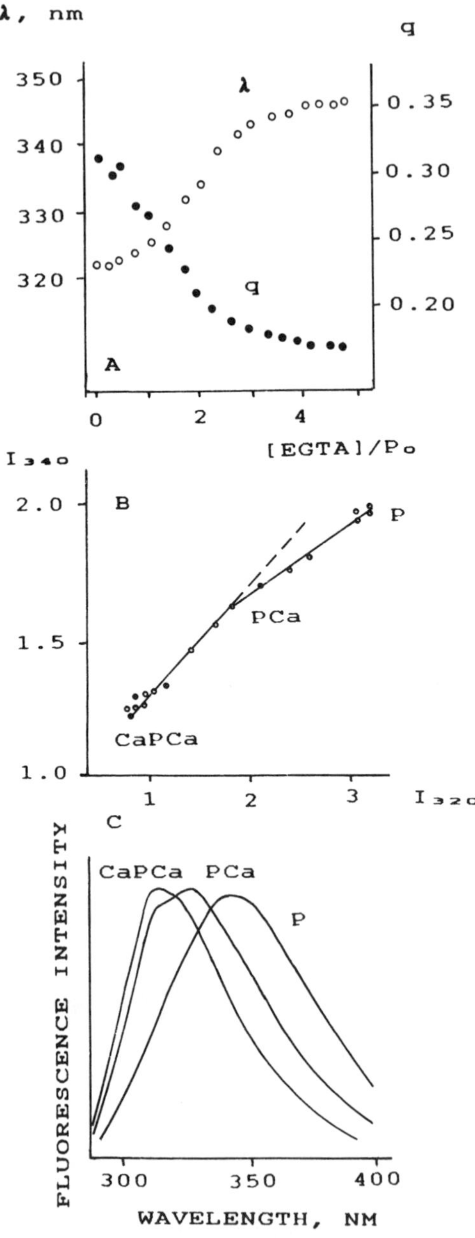

Figure 32. (A) Spectrofluorimetric EGTA titration of Ca^{2+}-loaded whiting parvalbumin (λ is fluorescence spectrum position; q is fluorescence quantum yield); (B) fluorescent phase plot; (C) fluorescence spectra of the initial, intermediate, and final states of parvalbumin.

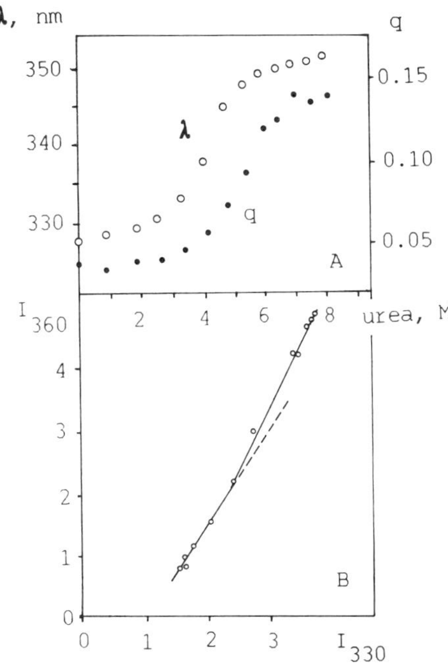

Figure 33. (A) Spectrofluorimetric urea titration of Ca^{2+}-loaded α-lactalbumin (λ is fluorescence spectrum position; q is fluorescence quantum yield); (B) fluorescent phase plot.

Construction of phase plots for a sufficient number of $I_{\lambda 1}$-$I_{\lambda 2}$ pairs and extrapolation of their linear parts up to the intersection point allow determination of the fluorescence spectra and yields of the intermediates.

VIII. STUDIES OF TEMPERATURE DEPENDENCIES OF PROTEIN FLUORESCENCE PARAMETERS

Investigation of the temperature dependence of protein fluorescence allows one to obtain information about the thermal region of stability of the native protein structure, and also about thermally induced changes within the limits of the native state.

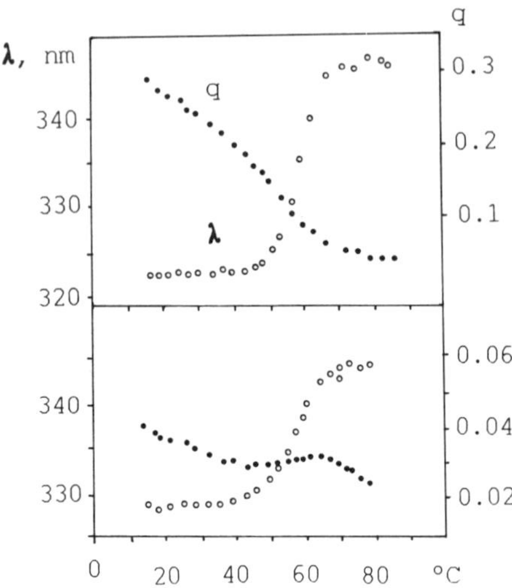

Figure 34. Temperature dependence of tryptophan fluorescence (○ — fluorescence spectrum position, λ, • — fluorescence quantum yield, q) of Mg^{2+}-loaded whiting parvalbumin (A) and Ca^{2+}-loaded bovine α-lactalbumin. (From Permyakov, E. A. and Burstein, E. A., *Biophys. Chem.*, 19, 265, 1984. With permission.)

Thermal denaturation of most tryptophan-containing proteins results, as a rule, in a more or less pronounced shift of the tryptophan fluorescence spectrum to longer wavelengths. It should be noted that the spectrum almost never shifts up to the position characteristic of tryptophan in completely aqueous environment.

Figure 34 shows the temperature dependencies of fluorescence parameters of whiting parvalbumin and bovine α-lactalbumin. In both cases the pronounced red spectral shift is caused by the thermal denaturation of these proteins.

In order to study thermally induced conformational changes in proteins, one should use fluorescence parameters linearly related to the fraction of conversion from native to denatured state α. As was mentioned above, such parameters are fluorescence quantum yield and intensity at a fixed wavelength. A problem in the use of these parameters in the study of the thermal transitions is that changes in them caused by the thermal unfolding of the protein structure develop against the background of the common temperature quenching induced by thermal activation of collisions of the excited chromophore with quenching groups of its environment. Fluorescence intensity at a fixed wavelength (or fluorescence quantum yield) at temperature T is related to the fraction of conversion α by the relationship

$$I_{\lambda,T} = (1-\alpha)\cdot I_{\lambda,TN} + \alpha \cdot I_{\lambda,TH} \qquad (15)$$

where indices N and H refer to the native and "high temperature" conformers, respectively, and $I_{\lambda,TN}$ and $I_{\lambda,TH}$ are the fluorescence intensities of these conformers at the temperature T.

In order to determine α, one should know how to measure $I_{\lambda,TN}$ and $I_{\lambda,TH}$ within the region of the thermal transition. The simplest way to evaluate them is a linear extrapolation of those parts of the temperature dependence $I_\lambda(T)$ corresponding to the native and denatured states of the protein to the thermal transition region. This method is rather rough because, for example, of the difficulties in determination of the temperature regions where the protein is in the native and denatured states.

Another method is based on the use of the fluorescent phase plots. Figure 35 shows such plots for parvalbumin and α-lactalbumin. Linear parts of the plots extrapolated to the origin correspond to common temperature fluorescence quenching without any changes in the protein structure. The complex curves between them correspond to the thermal transitions of the proteins. The beginning and the end of the transitions are well seen in the plots. $I_{\lambda,TN}$ and $I_{\lambda,TH}$ can be evaluated by extrapolation of the linear parts of the plots to the region of the thermal transitions. One should remember, however, that this is also a rather rough evaluation, since the values $I_{\lambda,TN}$ and $I_{\lambda,TH}$ on the extrapolated straight lines may not be on the vertical or horizontal straight line which passes through $I_{\lambda,T}$.

The most correct method of determination of $I_{\lambda,TN}$ and $I_{\lambda,TH}$ now seems to be the following. Bushueva et al.[18,19] showed that the temperature dependence of reciprocal fluorescence yield (or intensity at a fixed wavelength) for native proteins containing one emitting center within the region of nondenaturing temperatures is described by the equation

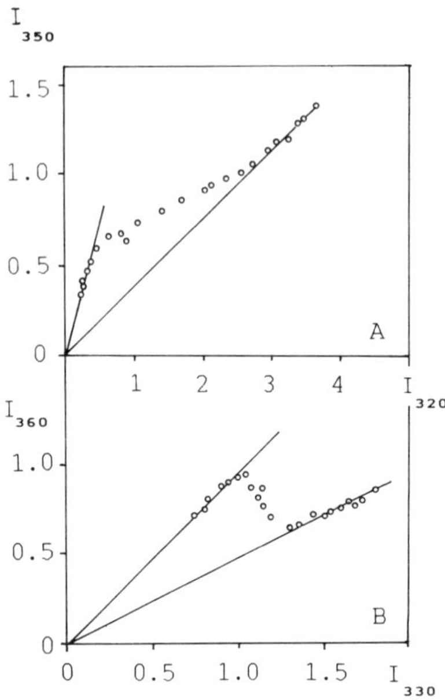

Figure 35. Fluorescence phase plots corresponding to the temperature dependence of fluorescence for Mg^{2+}-loaded whiting parvalbumin (A) and Ca^{2+}-loaded bovine α-lactalbumin (B). (From Permyakov, E. A. and Burstein, E. A., *Biophys. Chem.*, 19, 265, 1984. With permission.)

$$1/q = a + b \cdot T/\eta \tag{16}$$

where a and b are temperature-independent constants, T is temperature (K), and η is solvent viscosity (cP). The linearity of the 1/q (or 1/I) vs. T/η plots is thought to reflect the fact that the mobility of the internal parts of a protein globule is controlled by diffusion processes in the solvent. Though the strict linearity of the 1/q vs. T/η plots should be observed only for proteins with a single emitting chromophore, it is also observed for multitryptophan proteins. Figure 36 shows the 1/q vs. T/η plots for whiting parvalbumin and bovine α-lactalbumin.[20] Both plots have two linear parts corresponding to the proteins in the native and thermally denatured states. For elimination of the common thermal quenching effects, I_{λ,T^N} and I_{λ,T^H} values can be obtained

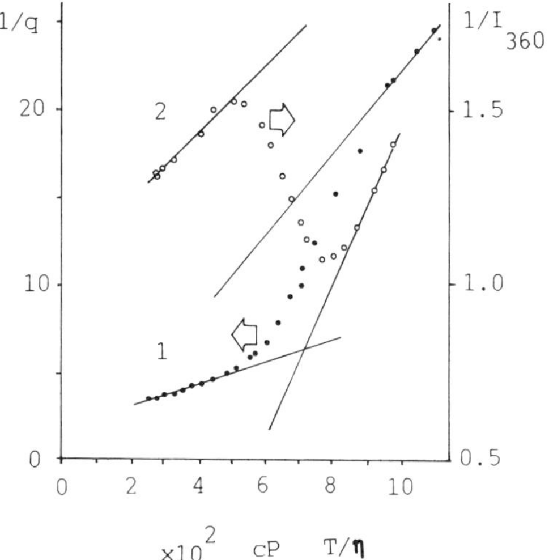

Figure 36. The 1/q (1/I) vs. T/η plots for whiting parvalbumin (1) and bovine α-lactalbumin (2). (From Permyakov, E. A. and Burstein, E. A., *Biophys. Chem.*, 19, 265, 1984. With permission.)

from extrapolations of the linear parts of the 1/I vs. T/η plots to the region of the thermal transition. Temperature dependencies of α obtained by this method for whiting parvalbumin and bovine α-lactalbumin are shown in Figure 37A. Figure 37B demonstrates the Van't-Hoff plots constructed from these data. The Van't-Hoff plot for α-lactalbumin is linear, but there is a break in that of whiting parvalbumin. This suggests the existence of at least two thermal transitions in parvalbumin. Using the plots presented in Figure 37, one can obtain thermodynamic parameters of the thermal transitions: enthalpy (ΔH), entropy (ΔS), and midtemperature (T_m) of the transition.

The higher temperature step of the thermal denaturation of whiting parvalbumin is accompanied by the red shift of the fluorescence spectrum by almost 20 nm. The spectral shift accompanying the lower temperature stage is much less pronounced (by 2 to 3 nm), i.e., the process of exposing of the tryptophan residue to the solvent mostly proceeds during the higher temperature stage of the thermal denaturation.

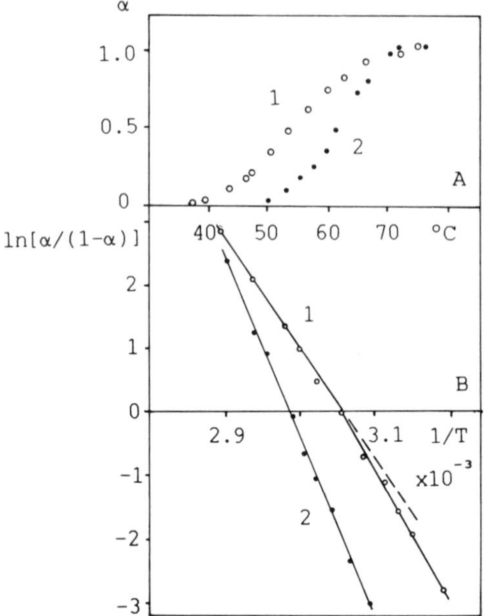

Figure 37. (A) Temperature dependence of parameter α for Mg^{2+}-loaded whiting parvalbumin (1) and Ca^{2+}-loaded bovine α-lactalbumin (2). (B) Van't-Hoff plots constructed from the data in A. (From Permyakov, E. A. and Burstein, E. A., *Biophys. Chem.*, 19, 265, 1984. With permission.)

This method of determination of α can be used even in cases in which the spectrum position or shape of the fluorescence spectrum do not change and the only informative parameter is fluorescence yield (or intensity). This is the case for tyrosine and phenylalanine fluorescence of proteins.

Figure 38 shows the temperature dependencies of the quantum yield of tyrosine fluorescence of pike parvalbumin, pI 4.2, and phenylalanine fluorescence of pike parvalbumin, pI 5.0. The position of their spectra remains constant within the temperature region from 10 to 85°C. Moreover, it is rather difficult to see a thermal transition in the curves.

Figure 39 shows the 1/q vs. T/η plots corresponding to these temperature dependencies. The curves between the linear parts

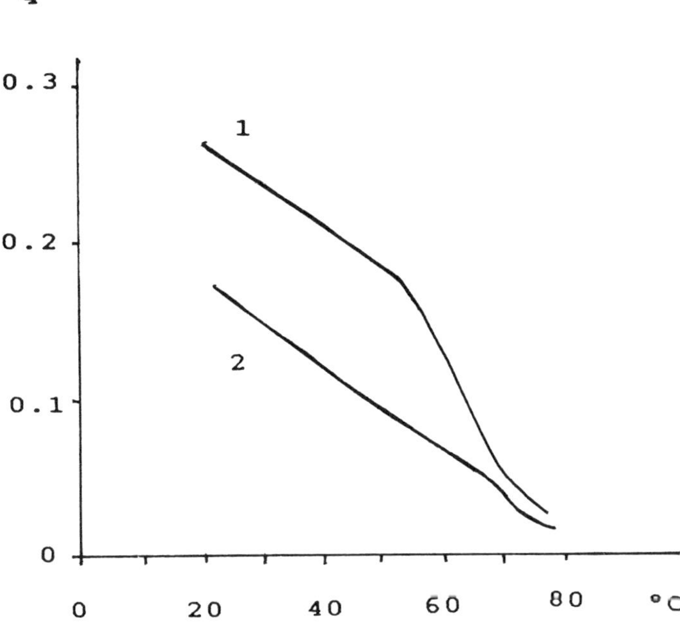

Figure 38. Temperature dependence of tyrosine fluorescence of Mg^{2+}-loaded pike parvalbumin, pI 4.2 (1), and phenylalanine fluorescence of Mg^{2+}-loaded pike parvalbumin, pI 5.0 (2). (From Permyakov, E. A. and Burstein, E. A., *Biophys. Chem.*, 19, 265, 1984. With permission.)

correspond to the thermally induced transitions in these proteins. Temperature dependencies of the fraction of conversion from the native to thermally denatured state, α, obtained by the method described above are represented in Figure 40. The curves have two stages. The same figure also shows calorimetric scans for these proteins. The maxima of the asymmetric heat sorption peaks correlate well with the midpoints of the higher temperature parts of the fluorescent curves, which shows that the main structural changes occur in this stage. The comparison of the fluorimetric and calorimetric data allow one to draw a conclusion about the correctness of the thermal denaturation curves obtained by the fluorescence method. An evaluation of thermodynamic parameters of the transitions carried out by these two methods gives similar results.

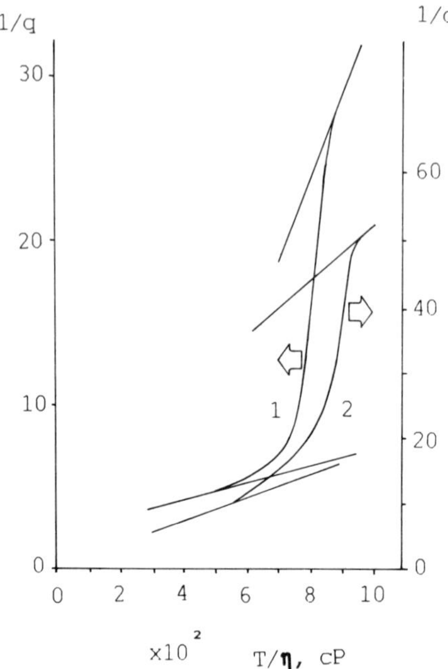

Figure 39. The 1/q vs. T/η plots for Mg^{2+}-loaded pike parvalbumins, pI 4.2 (1) and 5.0 (2). (From Permyakov, E. A. and Burstein, E. A., *Biophys. Chem.*, 19, 265, 1984. With permission.)

Some proteins possess thermal transitions within the limits of existence of the native structure. Calmodulin is an example of such a protein. Figure 41 shows the results of a study of the thermal transitions in bovine brain apocalmodulin carried out by fluorescence, microcalorimetry, and circular dichroism methods. Figure 41B shows temperature dependencies of the parameter α, obtained from the fluorimetric data by the method described above. It is clearly seen that the data of different methods agree well with each other. The heat sorption peak accompanied by a decrease in α-helical content and a decrease of fluorescence quantum yield corresponds to the thermal denaturation of the protein. There exists one more thermal transition in calmodulin at lower temperatures. This transition is detected both by fluorimetry and microcalorimetry.

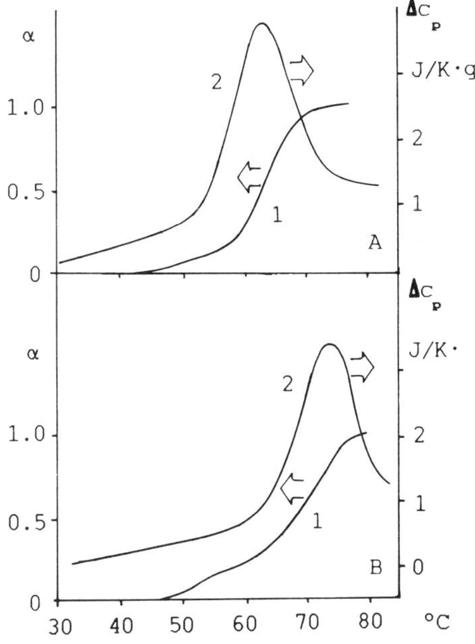

Figure 40. Temperature dependencies of parameter α (1) and partial specific heat capacity (2) for Mg^{2+}-loaded pike parvalbumins, pI 4.2 (A) and 5.0 (B). (From Permyakov, E. A. and Burstein, E. A., *Biophys. Chem.*, 19, 265, 1984. With permission.)

Table 2 contains the temperature dependence of water viscosity, which is useful in the analysis of the temperature dependencies of protein fluorescence parameters.[21]

IX. STUDIES OF PH DEPENDENCIES OF PROTEIN FLUORESCENCE

Studies of pH dependence of protein fluorescence allow determination of the region of protein native state in the pH scale and possible pH-induced conformational changes within the limits of the native structure. Changes in ionization state of protein groups can trigger conformational changes in proteins which, in turn,

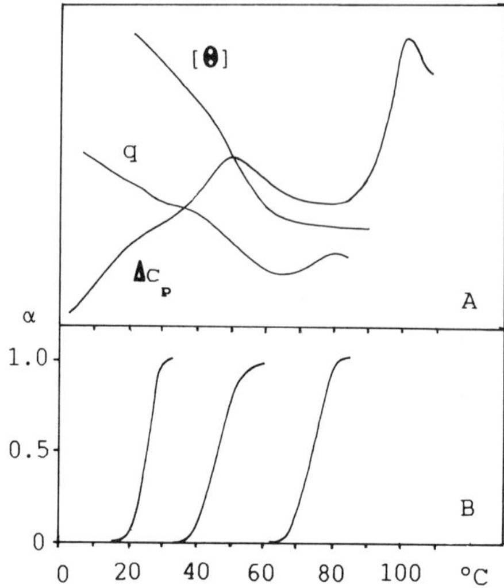

Figure 41. Thermal transitions in apocalmodulin. (A) Fluorescence quantum yield q; partial specific heat capacity C_p and ellipticity at 222 nm [θ]. (B) Temperature dependence of parameter α, obtained from the fluorimetric data. (From Permyakov, E. A., Shnyrov, V. L., et al., *Biochim. Biophys. Acta*, 830, 288, 1985. With permission.)

Table 2. Temperature Dependence of Water Viscosity

°C	η(cP)	°C	η(cP)	°C	η(cP)
0	1.7865	25	0.8909	70	0.4048
8	1.5138	30	0.7982	80	0.3554
10	1.3037	40	0.6540	90	0.3155
15	1.1369	50	0.5477	100	0.2829
20	1.0019	60	0.4674		

From Kaye, G. W. and Laby, T. H., *Tables of Physical and Chemical Constants*, Longmans, Green, London, 1959. With permission.

can change the environment of the protein chromophores and, hence, induce changes in protein fluorescence. There may also be another situation in which the changes in the ionization state of protein groups do not cause essential alterations in protein conformation but still change the protein fluorescence since they are located in the environment of the protein chromophores. In this case the fluorescence changes reflect changes in the quenching and relaxation properties of the nearest surrounding groups of the chromophores. One should be able to distinguish between these two situations.

In order to distinguish between fluorescence changes due to conformational changes and those due to changes in properties of the groups which are nearest to the chromophore, it is of importance to know which states of the ionogenic groups display stronger quenching properties. For example, one should know that in the case of tryptophan fluorescence, carboxylic groups and the imidazolic group of histidine are more effective quenchers when in the protonated state, while amino groups are stronger quenchers in the deprotonated state (see Chapter 3 of this book). Such data are not known for tyrosine and phenylalanine fluorescence. If changes of pH of a protein solution cause tryptophan fluorescence yield changes opposite to those expected from a knowledge of the quenching properties of groups titrated in this pH range, then these changes would seem to be caused by a pH-induced structural change. If fluorescence yield changes in the expected direction, then additional data are needed.

The changes of protein fluorescence at acidic pH values are caused by alteration of the ionic state of carboxylic groups of aspartic and glutamic acids. Figure 42 shows pH dependence of tryptophan fluorescence parameters for bovine α-lactalbumin.[10] The acidification of the solution results in the shift of the fluorescence spectrum from 327 to 341 nm and in the increase of fluorescence yield. Such spectral changes can be explained only by a pH-induced conformational change which transfers some of tryptophan residues from the protein interior to an aqueous environment. The existence of this conformational change, called the "acidic transition", is detected by many other physicochemical methods.

A specific feature of α-lactalbumin is its high affinity to Ca^{2+} ions. It is well known that the Ca^{2+} binding sites in proteins are formed by oxygen atoms of carboxylate and carbonyl groups. Upon the acidification of a protein solution, protons will compete with Ca^{2+} ions for carboxylic groups. At sufficiently acidic pH values all the carboxylate groups will be protonated, which will decrease the affinity of the protein to Ca^{2+}. Indeed, the spectral

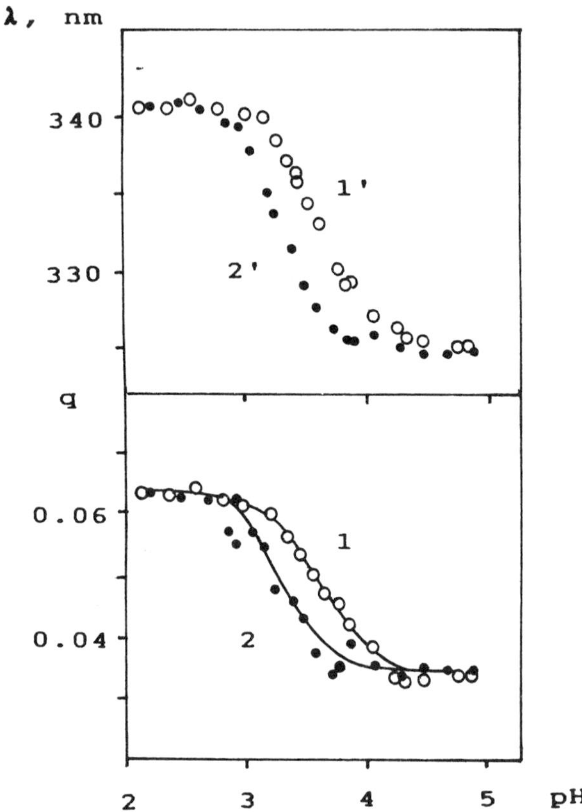

Figure 42. Spectrofluorimetric pH titration of α-lactalbumin. (A) Fluorescence spectrum position; (B) fluorescence quantum yield. (1) Ca^{2+} concentration equal to the protein concentration; (2) Ca^{2+} concentration 60 times higher than the protein concentration. In B the curves are theoretical. (From Permyakov, E. A., Yarmolenko, V. V., et al., *Biochem. Biophys. Res. Commun.*, 100, 191, 1981. With permission.)

changes induced by the acidification of the α-lactalbumin solution are similar to those caused by the Ca^{2+} release at neutral pH values.[10] An increase in Ca^{2+} concentration must shift the pH titration curves towards acidic pH values: protons do not compete easily with Ca^{2+}. This phenomenon is actually observed in experiment (Figure 42). The curves for pH dependencies of fluorescence quantum yield in Figure 42 are theoretical ones,

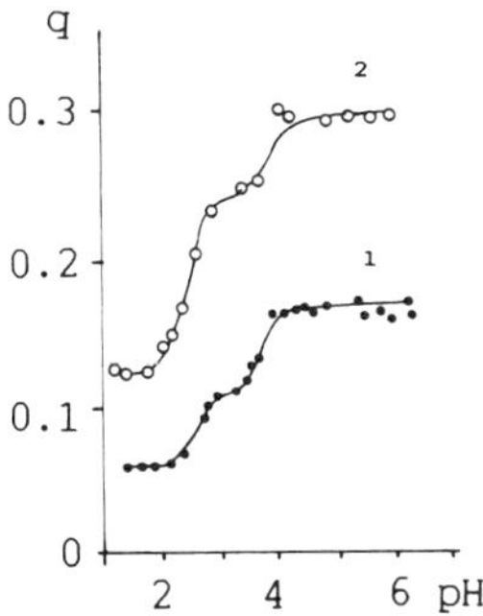

Figure 43. pH dependence of fluorescence quantum yield for pike parvalbumins, pI 5.0 (1) and 4.2 (2). (From Permyakov, E. A., Kalinichenko, L. P., et al., *Biochim. Biophys. Acta*, 749, 185, 1983. With permission.)

computed according to the scheme of competition of one Ca^{2+} ion and n H^+ ions for the same binding site:

$$\begin{aligned} P + Ca^{2+} &\overset{K}{\rightleftharpoons} PCa \\ P + nH^+ &\overset{pK}{\rightleftharpoons} PH_n \end{aligned} \qquad (17)$$

The curves were fit to the experimental points by variation of n and pK. The best fit was achieved when n = 3 and pK = 5.0, i.e., the Ca^{2+} binding site in α-lactalbumin contains three carboxylic groups with pK = 5. This example shows that the studies of pH dependence of protein fluorescence allow one to obtain valuable structural information.

Figure 43 shows pH dependencies of phenylalanine fluorescence of Ca^{2+}-loaded pike parvalbumin, pI 5.0, and tyrosine fluorescence of Ca^{2+}-loaded pike parvalbumin, pI 4.2, in the acidic pH region.[22] The decrease of fluorescence yield occurs in two stages, with midpoints at 3.7 and 2.6 for parvalbumin pI 5.0 and at 3.9 and 2.5 for parvalbumin pI 4.2. These spectral changes reflect the successive dissociation of the two bound Ca^{2+} ions

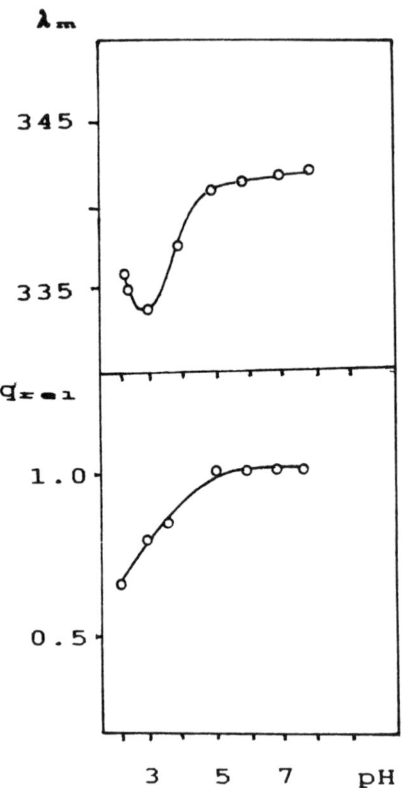

Figure 44. pH dependence of fluorescence spectral maximum position and fluorescence yield for human serum albumin. (From Iukova, M. N., Vedenkina, N. S., and Burstein, E. A., *Mol. Biol. (U.S.S.R.)*, 5, 214, 1971. With permission.)

induced by protonation of carboxylate groups taking part in the coordination of metal ions.

One of the classical objects of study of acidic transitions in proteins is serum albumin. The decrease of pH below 6 causes conformational changes in this protein which affect the environment of its tryptophan residues (human serum albumin has one tryptophan residue per molecule, while bovine serum albumin has two tryptophan residues). Figure 44 shows the pH dependence of fluorescence spectral maximum position and fluorescence yield for human serum albumin.[23] In the pH region from

5 to 3, the spectrum maximum shifts from 341.5 to 333 nm. The spectral changes correspond to the so-called N → F transition in serum albumin, detected by other methods as well. The F-form is characterized by a looser structure with disordered hydrophobic regions and disrupted salt bridges. In the course of this structural change, the tryptophan residue is transferred from the aqueous environment on the protein surface to a hydrophobic interior of the protein. At pH <3 there exists one more structural rearrangement, the acidic expansion, which is reflected in a red shift of the emission spectrum, i.e., the tryptophan residue becomes more accessible to water.

At alkaline pH values, fluorescence changes are usually caused by deprotonation of such groups as tyrosine, lysine, and arginine residues. One of the difficulties in studies of pH dependencies of fluorescence for tyrosine-containing proteins at alkaline pH is consideration of the appearance in their absorption spectra of tyrosinate bands. Even if we detect only tryptophan fluorescence, the tyrosinate absorption should be taken into account. Since the long-wavelength wing of the tyrosinate absorption spectrum spreads over 300 nm, we should correct the tryptophan emission spectrum for the reabsorption by tyrosinates of a part of the light emitted by tryptophan residues. Moreover, if the fluorescence is excited by the light with a wavelength below or above 280 nm, we should correct the fluorescence spectrum for the screening effect, i.e., we should take into account that a part of the excitation light will be absorbed by tyrosinates. At 280 nm the absorption of tyrosine and tyrosinate is the same; therefore, upon the excitation at 280 nm the correction for the screening effect is not needed. Methods of practical correction of fluorescence spectra for screening and reabsorption effects are in Chapter 2.

Figure 45 shows the pH dependence of tryptophan fluorescence parameters of α-lactalbumin in various ionic states: in the apo-state and in Ca^{2+}-, Na^{+}-, and K^{+}-loaded states.[24] The red shift of the fluorescence spectrum is caused by alkaline denaturation of the protein. The appearance of the pH dependence of fluorescence yield suggests the existence of an intermediate state in the course of the alkaline denaturation, which is corroborated by fluorescent phase plots. Deprotonation of the tyrosine side chains correlates with the second stage of the spectral changes. The structural changes in this pH region seem to be triggered by

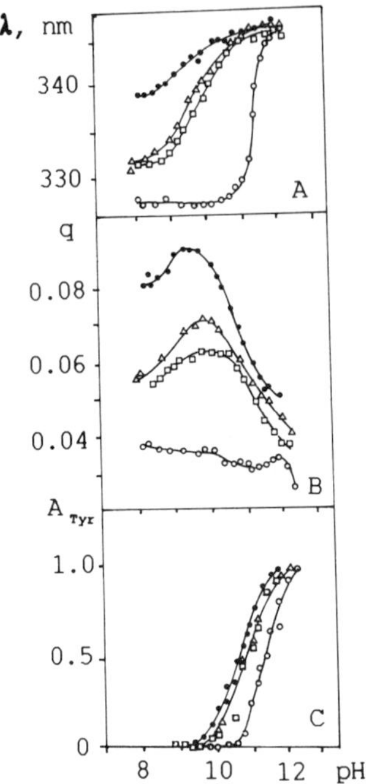

Figure 45. pH dependencies of tryptophan fluorescence spectrum position (A) and quantum yield (B) and normalized absorption of tyrosinate at 300 nm (C) for different ionic states of α-lactalbumin: ○ — Ca^{2+} loaded; Δ — Na^{+} loaded, □ — K^{+} loaded; · — apo-state.

ionization of the tyrosine residues. The conformational changes in the more acidic region seem to be caused by deprotonation of lysine side chains. The Ca^{2+}-loaded state of α-lactalbumin is the most stable one with respect to alkaline denaturation, while the apo-state is the least stable one.

Figure 46 shows the pH dependence of quantum yield of tyrosine fluorescence and normalized absorption of tyrosinate at 300 nm for bovine brain calmodulin (2 Tyr, 0 Trp) in Ca^{2+}-loaded and apo-states.[25] The curves for pH dependence of emission quantum yield are corrected for the effects of tyrosinate screening and reabsorption. The dependence, q(pH), is approximated well by a one-proton titration curve with $pK_a = 10.7$, while the curves for normalized tyrosinate absorption are more sloping and can be approximated only by two successive one-proton

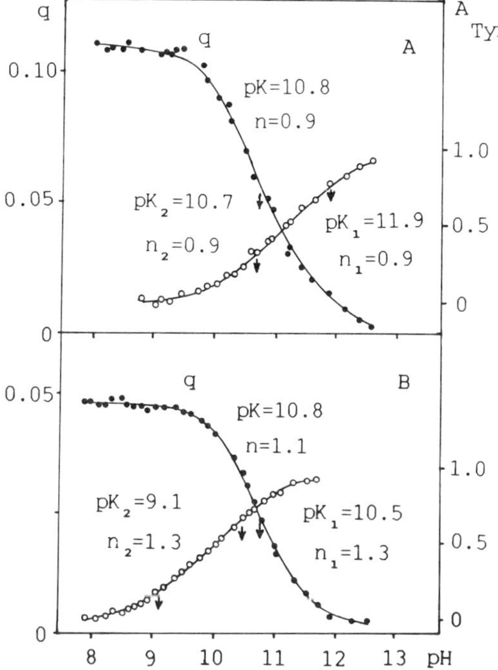

Figure 46. pH dependencies of tyrosine fluorescence quantum yield q and normalized tyrosinate absorption at 300 nm for bovine brain calmodulin. (A) Ca^{2+} loaded; (B) apo-state. pK and n are determined from the fit according to Scheme 17. (From Permyakov, E. A., Shnyrov, V. L., et al., *Biochim. Biophys. Acta,* 830, 288, 1985. With permission.)

curves. Values of pK_a and n were determined by means of the fitting of theoretical curves computed according to the scheme

$$P + nH^+ \underset{}{\overset{pK_a}{\rightleftharpoons}} PH_n \tag{18}$$

to experimental points by variation of Pk_a and n. It turned out that the parameters of the dependence, q(pH), do not depend on Ca^{2+} concentration, while the pH dependence of tyrosinate absorption changes its position depending on Ca^{2+} concentration. However, one of the one-proton curves conserves practically

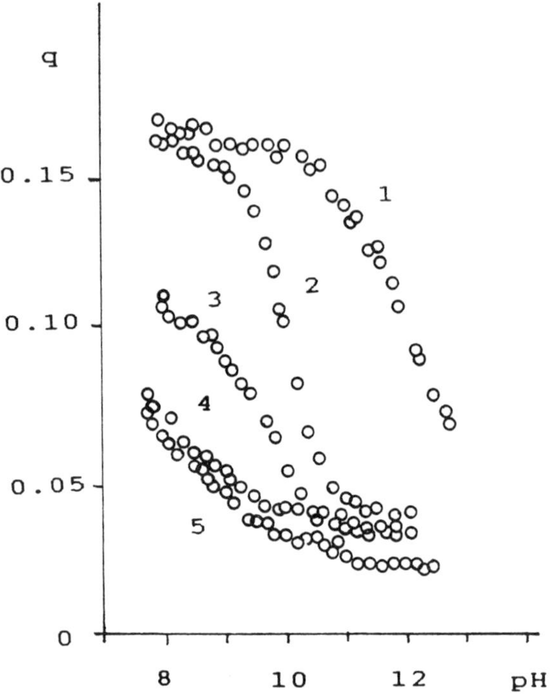

Figure 47. pH dependencies of phenylalanine fluorescence quantum yield of pike parvalbumin, pI 5.0, in Ca^{2+}-loaded (1), Mg^{2+}-loaded (2), Na^{+}-loaded (3), K^{+}-loaded (4) and apo-states (5). (From Permyakov, E. A., Kalinichenko, L. P., et al., *Biochim. Biophys. Acta*, 749, 185, 1983. With permission.)

the same pK_a and n as the q(pH) curves. It is reasonable to assume that only one tyrosine residue in calmodulin emits. The pK_a and n values of this tyrosine residue do not depend on Ca^{2+} concentration. Another tyrosine residue almost does not emit, but its pK_a is Ca^{2+} sensitive. This example demonstrates that the study of pH dependence of tyrosine fluorescence and tyrosinate absorption can also provide rather interesting information about protein.

Less rich information can be obtained in the case of phenylalanine fluorescence. Figure 47 shows the pH dependence of phenylalanine fluorescence of pike parvalbumin, pI 5.0, for Ca^{2+}-, Mg^{2+}-, Na^{+}-, and K^{+}-loaded states and also for the apo-protein.

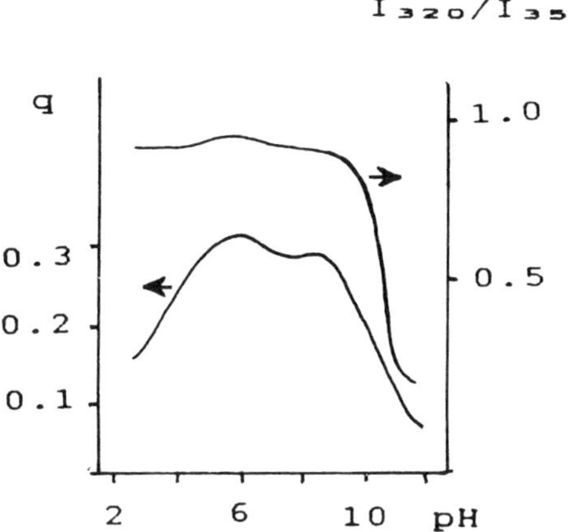

Figure 48. pH dependence of fluorescence quantum yield q and a value characterizing spectrum position I_{320}/I_{350} for RNAase C2. (From Grishchenko, V. M., Emelyanenko, V. I., et al., *Bioorg. Chem.*, 2, 207, 1976. With permission.)

The higher the affinity of a metal ion to the protein, the stronger its stabilization of the protein.

Within the neutral pH region, the changes of protein fluorescence are usually caused by changes in the ionization state of either the N terminal amino group of the imidazole of histidine residues. Figure 48 shows the pH dependence of fluorescence quantum yield and of a value which characterizes spectrum position, i.e., the ratio of fluorescence intensities at short- and long-wavelength wings of the spectrum, I_{320}/I_{350}, for RNAase C2 *Aspergillus clavatus*.[7] The decrease of pH from 7.5 to 6.0 causes a pronounced (up to 30%) increase in fluorescence yield accompanied by a small (up to 0.5 nm) blue spectral shift. These changes are caused by protonation of the imidazole ring of one or two of the histidines ($pK_a = 6.5$) of RNAase C2. The increase in fluorescence yield and the blue shift in the spectrum suggest an increase in the rigidity of tryptophan environment and elimination of the quenching action of some neighboring groups. Since imidazole in cationic form is a much stronger quencher than in neutral form, it is reasonable to conclude that the change in the charge of

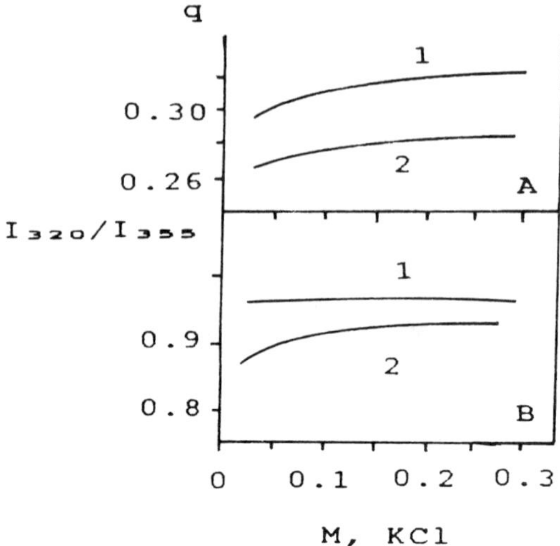

Figure 49. Effects of KCl on fluorescence quantum yield (A) and spectrum position (B) of RNAase C2 at pH 6.0 (1) and 8.0 (2). (From Grishchenko, V. M., Emelyanenko, V. I., et al., *Bioorg. Chem.*, 2, 207, 1976. With permission.)

one or two histidine residues causes a conformational change which is reflected in the fluorescence changes.

X. STUDIES OF EFFECTS OF IONIC STRENGTH ON PROTEIN FLUORESCENCE

One more universal factor that can affect protein conformation is ionic strength. An increase in ionic strength, i.e., ion concentration in solution, causes a screening of protein-charged groups by surrounding counterions. This weakens electrostatic interactions between protein groups and can change protein conformation. As a rule, spectral changes induced by ionic strength changes are smooth.

Figure 49 shows dependencies of fluorescence parameters of RNAse C2 on KCl concentration measured at two different pH values.[7] The increase in KCl concentration from 0.025 to 0.15 *M* causes an increase in fluorescence yield by 7 to 9% at both pH values. Besides that, at pH 8.0 the increase of the KCl concentration (up to 0.3 *M*) induces a blue shift of the fluorescence spectrum. Further increase in KCl concentration (up to 0.3 *M*) essentially does not influence the fluorescence parameters. The observed increase in fluorescence yield is caused by local changes in indole chromophore environment, while the blue spectral shift at pH 8 is explained by a decrease of the mobility of polar groups. Perhaps the region surrounding the tryptophan residue of RNAase C2 is more rigid at pH 6 than at pH 8. The observed effects are caused by a screening of protein-charged groups.

In some cases, one should distinguish the nonspecific effects of ionic strength from the effects of ion binding. One should be especially careful in the case of metal-binding proteins.

Figure 50 shows dependencies of fluorescence parameters of whiting parvalbumin on NaCl and KCl concentration.[26] The increase of NaCl and KCl concentration causes a blue spectral shift and an increase in the fluorescence yield. It is of importance that NaCl and KCl induce quantitatively different changes, which demonstrates that in this case we are not observing the effect of ionic strength, but the effects of Na^+ and K^+ ions binding to the parvalbumin. Since in the case of Na^+ the curves plateau at lower concentrations, one can conclude that the affinity of the protein to Na^+ is higher than that to K^+.

XI. STUDIES OF EFFECTS OF DENATURANTS ON PROTEIN FLUORESCENCE

One of the methods of study of structural features of proteins is the study of their denaturation induced by such compounds as urea or guanidine hydrochloride. The denaturation of many proteins occurs only at very high concentrations of these compounds. As a rule, both urea and guanidine hydrochloride denature proteins practically up to the random coil state. At high concentrations of these denaturants the fluorescence spectrum position for many proteins coincides with that for free tryptophan in water.

Figure 51 shows the dependence of fluorescence parameters of the single tryptophan residue in whiting parvalbumin on urea

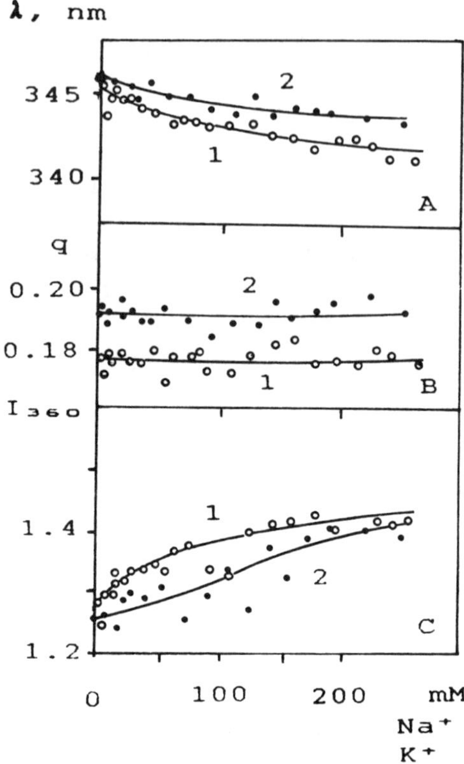

Figure 50. Dependence of fluorescence spectrum position (A), quantum yield (B), and fluorescence intensity at 360 nm (C) of whiting parvalbumin on NaCl (1) and KCl (2) concentration. (From Permyakov, E. A., Kalinichenko, L. P., et al., *Biochim. Biophys. Acta*, 749, 185, 1983. With permission.)

concentration.[27] The increase in urea concentration causes very pronounced spectral changes. In 7.5 *M* urea, fluorescence spectrum is red shifted by 33 to 34 nm, compared with the spectrum of the native protein, and the position of its maximum coincides with that of free tryptophan in water. The main part of the spectral shift occurs within urea concentrations from 4 to 6 *M*. Nevertheless, an essential change of fluorescence yield accompanied by a small red shift and widening of the spectrum takes place at lower urea concentrations. The two-stepped decrease of fluorescence yield suggests an intermediate state of the protein. A

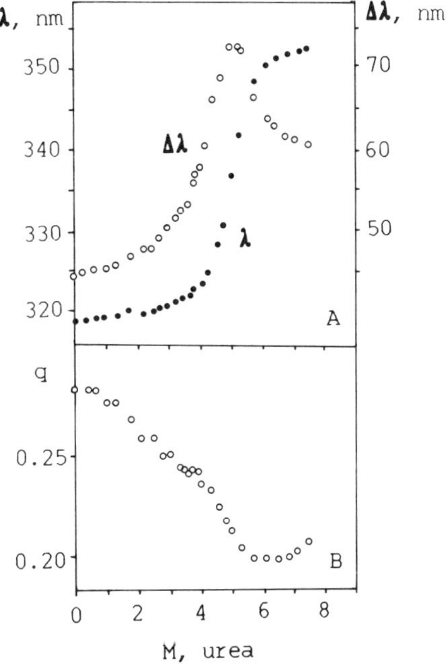

Figure 51. Dependence of fluorescence parameters of whiting parvalbumin on urea concentration. (A) spectrum position λ, spectrum width Δλ; (B) fluorescence quantum yield. (From Permyakov, E. A., Kalinichenko, L. P. et al., *Biochim. Biophys. Acta,* 749, 185, 1983. With permission.)

fluorescence phase plot (Figure 52) confirms the existence of the intermediate state in about 3.2 *M* urea. The fluorescence spectrum of the protein in 3.2 *M* urea is shown in Figure 52C. It is close in position and shape to the spectrum of the protein with one bound Ca^{2+} ion. One can believe that in the first stage of the denaturation process the protein structure does not change very much, but that Ca^{2+} affinity of the weaker binding site decreases and the protein loses one of the bound Ca^{2+} ions.

There is one more change of tryptophan environment within the urea concentration range from 6 to 7.7. This is reflected in a gradual distortion of the phase plot (Figure 52) and in an increase in fluorescence quantum yield (Figure 51B). The nature of these changes is not clear at present. It may be that they are caused by an additional binding of urea to the denatured protein.

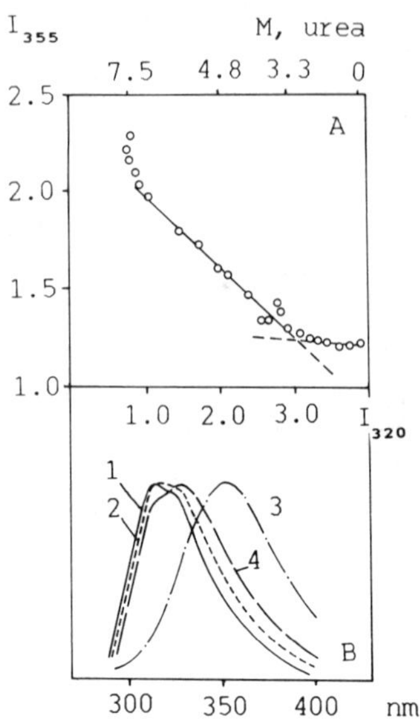

Figure 52. (A) Fluorescent phase plot corresponding to the urea-induced denaturation of whiting parvalbumin (Figure 51). (B) Fluorescence spectra of the parvalbumin at different urea concentrations: (1) in the absence of urea; (2) 3.2 *M* urea; (3) 7.5 *M* urea. (4) Fluorescence spectrum of the protein with one bound Ca^{2+}. (From Permyakov, E. A., Kalinichenko, L. P., *Biochem. Biophys. Acta*, 749, 185, 1983. With permission.)

It should be noted that for metal-binding proteins the denaturing action of urea or guanidine hydrochloride strongly depends on metal ion concentrations. The greater the ion concentration, the more difficult it is to denature the protein. In some cases, addition of a sufficiently high quantity of ions to the protein in the presence of extremely high concentrations of denaturants may practically return it to the native state. Figure 53 shows the results of Ca^{2+} titration of whiting parvalbumin in the presence of 7.5 *M* urea. The increase in Ca^{2+} concentration causes changes of fluorescence parameters which are opposite to those induced by the increase of urea concentration. The plots reach a plateau at a relative Ca^{2+} concentration $[Ca^{2+}]/[\text{protein}] \approx 100$.

As was shown earlier, one of the methods for detection of intermediate states of a protein that arise during the course of the

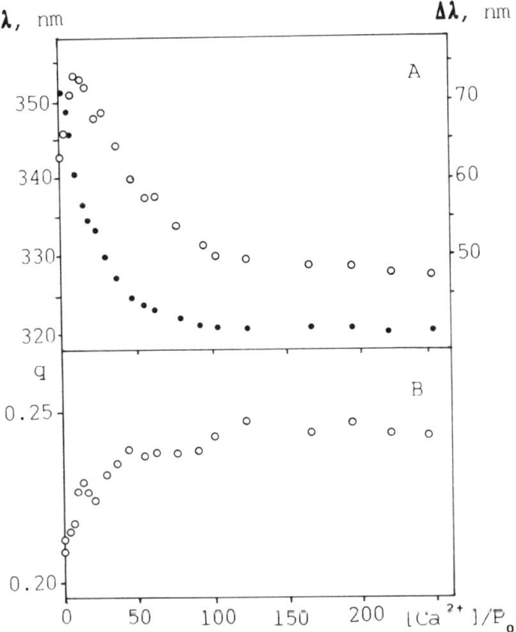

Figure 53. Ca^{2+} titration of whiting parvalbumin in the presence of 7.5 *M* urea. (A) Spectrum position and width; (B) fluorescence yield. (From Permyakov, E. A., Kalinichenko, L. P., *Biochem. Biophys. Acta,* 749, 185, 1983. With permission.)

denaturation process is the construction of the fluorescent phase plots. In some cases, this detection cannot be done without phase plots. Figure 33 shows fluorescence data for urea denaturation of Ca^{2+}-loaded α-lactalbumin.[24] No protein intermediate states can be observed from the plots in Figure 33A, while the fluorescent phase plot in Figure 33B clearly demonstrates the existence of an intermediate state.

Tyrosine and phenylalanine fluorescence can also be used for studies of protein denaturation. Figure 54 shows the dependencies of the quantum yields of tyrosine and phenylalanine fluorescence of Ca^{2+}-loaded pike parvalbumins, pI 4.2 and 5.0, on urea concentration. The denaturation-induced increase in fluorescence yield begins at 7.5 *M* for parvalbumin pI 5.0 (phenylalanine fluorescence) and at 6.0 *M* for parvalbumin pI 4.2 (tyrosine fluorescence). The information obtained in the case of phenylalanine

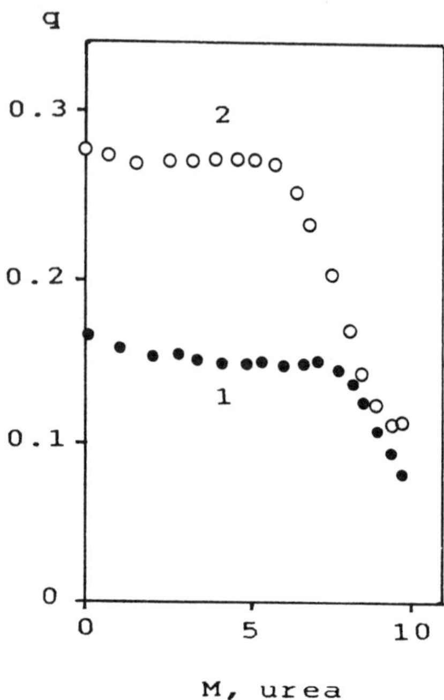

Figure 54. Dependence of fluorescence quantum yield of pike parvalbumins, pI 5.0 (1; phenylalanine fluorescence) and pI 4.2 (2; tyrosine fluorescence), on urea concentration. (From Permyakov, E. A., Medvedkin, V. N., et al., *Arch. Biochem. Biophys.*, 227, 9, 1983. With permission.)

and tyrosine fluorescence is much poorer compared with that provided by tryptophan emission.

XII. STUDIES OF INTERACTIONS OF PROTEINS WITH LOW-MOLECULAR-MASS COMPOUNDS BY FLUORESCENCE METHOD

Many proteins are able to interact with some low-molecular-mass substances such as metal ions and various organic compounds: nucleotides, substrates, products, cofactors of reactions, and so on. In some cases, these interactions can be studied by means of the intrinsic fluorescence method. The main condition for the possibility of such studies is the existence of a measurable fluorescent signal on the binding of a compound to the protein. However, there are possible situations in which a protein binds a compound, but its intrinsic fluorescence does not reflect this, i.e., the changes in protein conformation induced by the binding do not affect the environment of the protein chromophores. In the case of proteins containing many tryptophan residues, there may be a situation in which the binding of a compound changes the environment of a small number of the chromophores, which would be practically invisible against the background of a strong emission of other chromophores. So the absence of a fluorescent signal on the addition of a substance to the protein solution cannot be taken as a proof of the absence of interactions between the protein and the substance.

The determination of parameters of interactions of a protein with a compound by the fluorescence method is a rather difficult problem. Let's consider some practical situations.

Let a protein P bind a compound S with binding constant K, n molecules of S being bound per protein molecule. It should be noted that usually it is hard to determine n from fluorescent data. The simplest case is when n = 1:

$$P + S \underset{}{\overset{K}{\rightleftharpoons}} PS \tag{19}$$

In order to have an opportunity to study this reaction by the fluorescence method, the fluorescence parameters of P and PS should be different.

Usually, in order to study the binding of S to the protein, the protein solution is titrated by small additions of S, and this process is monitored by intrinsic protein fluorescence. Fluorescence yield (or intensity at a fixed wavelength) will change according to the equation

$$q = q_p \cdot [P]/P_0 + q_{PS} \cdot [PS]/P_0 \tag{20}$$

where q_p and q_{ps} are fluorescence yields of P and PS, while P_0 is total protein concentration. The equations for mass balance

$$[PS] + [P] = P_0 \tag{21}$$

$$[PS]+[S]=S_0 \tag{22}$$

(S_0 is total concentration of S) and the definition of binding constant

$$K=\frac{[PS]}{[P]\cdot[S]} \tag{23}$$

allow one to obtain the dependence of [P] and [PS] on S_0 and P_0:

$$[P]=\frac{P_0}{1+K\cdot[S]} \tag{24}$$

$$[PS]=\frac{K\cdot[S]\cdot P_0}{1+K\cdot[S]} \tag{25}$$

where

$$[S]=\frac{\left\{\left[1+K(P_0-S_0)\right]^2+4KS_0\right\}^{1/2}-\left[1+K(P_0-S_0)\right]}{2K} \tag{26}$$

The appearance of the curve $q(S_0)$ will depend on the relationship between protein concentration P_0 and dissociation constant 1/K.

If $P_0 >> 1/K$, the dependence $q(S_0)$ will be a straight line with a break at $S_0 = P_0$. If $P_0 = 1/K$, then at $S_0 = P_0$, $q - q_p = 0.38$ ($q_{ps} - q_p$), i.e., the $q(S_0)$ will be gently sloping. Upon still lower concentrations of the protein, the curve will be still more sloping (Figure 55).

It is evident that it is impossible to determine the binding constant, K, from the dependence of $q(S_0)$ when $P_0 >> 1/K$. However, binding stoichiometry is clearly seen in this case. On the other hand, when $P_0 << 1/K$, it is difficult to measure stoichiometry, while the binding constant can be determined with sufficient accuracy.

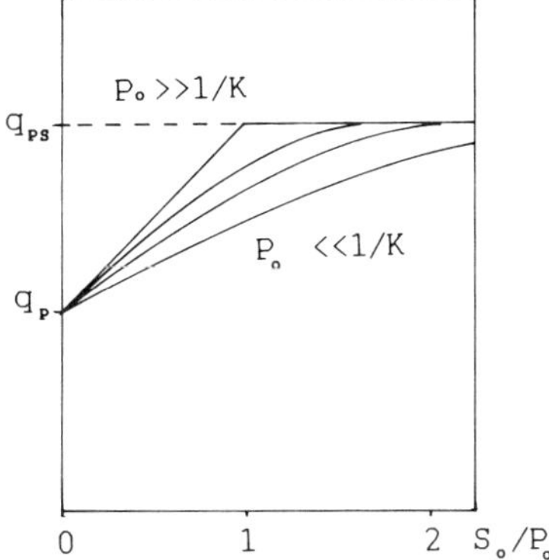

Figure 55. Theoretical curves for dependence of protein fluorescence quantum yield on relative total concentration of a compound S at different protein concentrations.

In the case of a high value of K, a decrease in P_0 in order for it to be comparable with 1/K is often impossible because of poor sensitivity of the spectral instrument. Therefore, we are forced to carry out measurements under conditions in which $P_0 >> 1/K$ and the binding constant value cannot be determined.

In the opposite case of a very low binding constant, K, its value can be determined easily if the binding stoichiometry is known. In this case, the dependence $q(S_0)$ reaches plateau at very high S_0 concentrations when $[S] \approx S_0$.

As was mentioned above, the best situation is when P_0 is comparable to 1/K. In this case, both binding constant and stoichiometry can be determined. When n = 1, one can think that the value

$$\alpha = \frac{q - q_P}{q_{PS} - q_P} \tag{27}$$

equals $[PS]/P_0$, i.e.,

$$\alpha = \frac{K \cdot [S]}{1 + K \cdot [S]} \tag{28}$$

or in the Scatchard's coordinates,[28,29]

$$\alpha = 1 - \frac{1}{K} \cdot \frac{\alpha}{[S]} \tag{29}$$

[S] can be evaluated from experimental data:

$$[S] = S_0 - [PS] = S_0 - \alpha \cdot P_0 \tag{30}$$

K can be determined from the slope of the Scatchard plot (the dependence of $\alpha/[S]$ on α).

A more convenient and more accurate method for determination of K is a computer fitting of the theoretical dependence $q(S_0)$ to experimental points by variation of the value of K.

If a protein possesses two binding sites for S, many additional problems arise, and first of all is the problem of the binding mechanism. The sites may either be independent or there may be interactions between them in which the binding of S to one of them influences the affinity of the other (binding cooperativity). As a rule, it is hard to recognize these cases using only the intrinsic fluorescence method.

If it is known that the binding sites are filled successively, which can be either a case of independent binding when the binding constants for the first and the second sites essentially differ, or a case of cooperative binding when the binding to the second site becomes possible only after the filling of the first site, one should know what fluorescent signal corresponds to the filling of each of the sites. It is better to determine it when $P_0 >> 1/K$. The curve $q(S_0)$ may have a different character (Figure 56). Most often the fluorescent response to the filling of the first site is the largest; the filling of the second site usually gives the same or smaller signal. It should be noted that, in principle, it is possible to have a situation in which the filling of the first site causes a smaller signal than the filling of the second one. The $q(S_0)$ curve can be sigmoidal in this case, which might be considered to be a reflection of cooperative interactions between the sites which are actually absent.

The determination of binding constants should be carried out at protein concentrations comparable to the dissociation constants

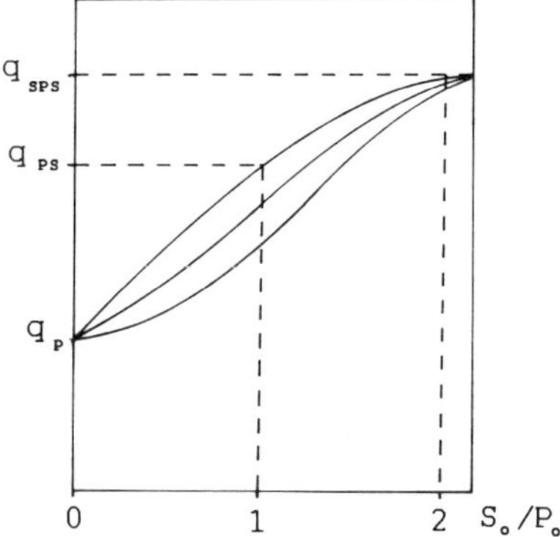

Figure 56. Theoretical curves for the dependence of protein fluorescence quantum yield on relative total concentration of S when the protein possesses two S binding sites for different relationships between the fluorescence yields of P, PS, and SPS.

K_1 and K_2. The most convenient method for this is the computer fitting of the theoretical curve computed according to a suggested binding scheme to experimental points by variation of K_1 and K_2.

If a protein has still more binding sites, the interpretation of fluorescent data becomes still more complicated.

Let's consider some concrete examples of studies of interactions of proteins with low-molecular-mass substances by the fluorescence method.

Figure 57A shows the dependence of fluorescence parameters of ion-free bovine α-lactalbumin on Ca^{2+} concentration.[10] The binding of Ca^{2+} to the protein causes a blue shift of the fluorescence spectrum and a decrease of fluorescence quantum yield. Only fluorescence parameters which are linearly related to the fraction of conversion can be used for quantitative evaluations. As was mentioned above, such a parameter is fluorescence quantum yield q. It is clearly seen from Figure 57A that the dependence $q([Ca^{2+}]/P_0)$ is a straight line with a break at $[Ca^{2+}]/P_0 = 1$. It means that fluorescence feels the binding of a single Ca^{2+} per molecule and that the experiment was carried out in conditions such that $P_0 >> 1/K$. Due to the limited sensitivity of the instrument, we cannot decrease protein concentration down to

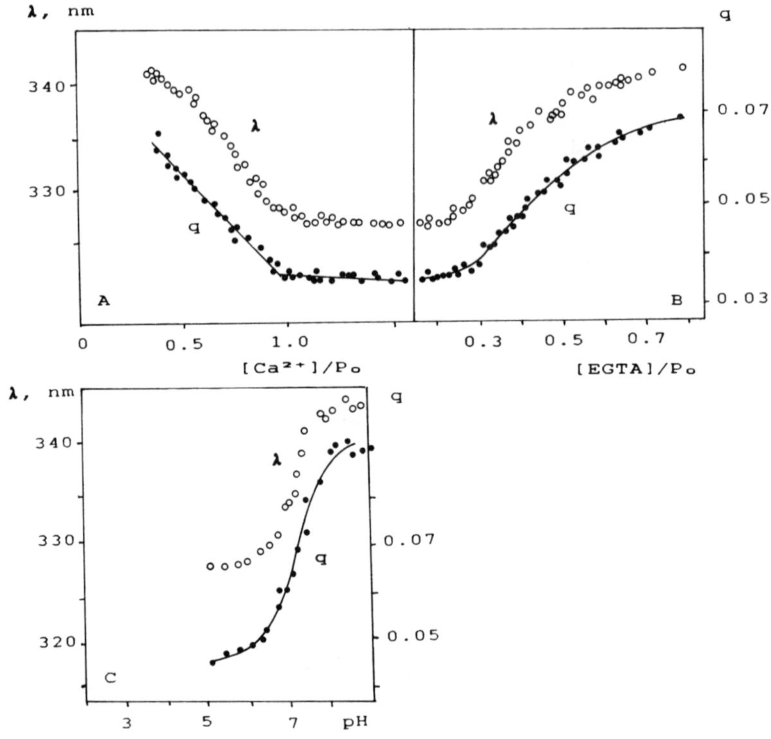

Figure 57. Spectrofluorimetric Ca^{2+} (A), EGTA (B) pH 8.0, 46 μ*M* Ca^{2+}), and pH (C) titration of bovine α-lactalbumin. Protein concentration P_0 = 30 μ*M*.

the level where it would be comparable to 1/K, so we cannot determine the binding constant from the direct Ca^{2+} titration experiment.

Nevertheless, in the case of metal cations, the measurement of the binding constant by the fluorescence method is still possible, but for this we should use a sufficiently strong chelator of divalent cations which was known cation-binding properties. For Ca^{2+} ions we can use EGTA and EDTA.

Figure 57B shows results of spectrofluorimetric titration of Ca^{2+}-loaded α-lactalbumin by the chelator EGTA. It is clearly seen that the increase in EGTA concentration causes changes which are opposite to those induced by the Ca^{2+} binding. An essential condition of such an experiment is that the effective Ca^{2+} binding constant of the chelator should be comparable to the Ca^{2+} binding constant of the protein — only in this case, from data obtained for competition of the protein and chelator for calcium ions, can we evaluate the Ca^{2+} binding constant with sufficient accuracy. If the Ca^{2+} binding constants of the protein and chelator are comparable with each other, the curve of the spectrofluorimetric EGTA titration is smooth and gently sloping. If the chelator has a much higher affinity to Ca^{2+} in comparison with the protein, the curve of the spectrofluorimetric EGTA titration almost does not differ from a straight line, and in this case the determination of the binding constant is also impossible. In the opposite case, when the chelator binds Ca^{2+} much more weakly than does the protein, the curve of the spectrofluorimetric EGTA titration has too steep a slope, and we would have to add too much chelator to remove calcium ions from the protein. In many cases this is undesirable, since some proteins are able to bind EGTA and EDTA, and though this binding is usually weak, it can distort the results of the experiment at high concentrations of the chelators.

By changing pH we can change the effective Ca^{2+} binding constant of EGTA and EDTA. Figure 58 shows the pH dependencies of logarithms of Ca^{2+} and Mg^{2+} constants of EGTA and EDTA.[30] If the properties of a protein allow it, one can make the effective Ca^{2+} binding constant of EGTA comparable with the Ca^{2+} binding constant for the protein by changing pH.

The simplest scheme of competition of EGTA and the protein for calcium ions is the following one:

$$\begin{aligned} P + Ca^{2+} &\underset{}{\overset{K}{\rightleftharpoons}} PCa \\ E + Ca^{2+} &\underset{}{\overset{K_E}{\rightleftharpoons}} ECa \end{aligned} \qquad (31)$$

where P and E are the protein and EGTA, respectively. Fitting of a theoretical curve computed according to Scheme 31 to the experimental points by variation of K allows evaluation of the K value. For α-lactalbumin, the Ca^{2+} binding constant determined by this method is $3 \times 10^8\ M^{-1}$ at 20°C.

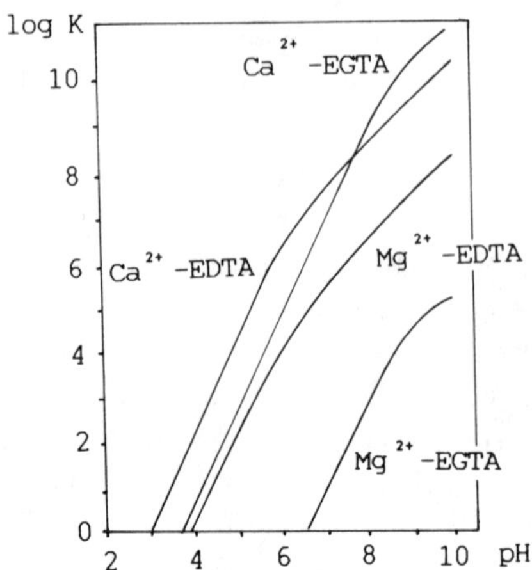

Figure 58. pH-dependence of logarithms of effective Ca^{2+} and Mg^{2+} binding constants for EGTA and EDTA. (From Schwarzenbach, G. and Flaschka, H., *Die Komplexonometrische Titration,* Ferdinand Enke Verlag, Stuttgart, 1965. With permission.)

In order to evaluate the Ca^{2+} binding constant for a Ca^{2+}-binding protein, one can use another method. Since Ca^{2+} ions are coordinated in proteins by oxygen atoms of carbonyl and carboxyl groups, usually Ca^{2+}-binding properties of the proteins do not depend on pH for the pH region above 6. In contrast to this, effective association constants of the Ca^{2+}-EGTA and Ca^{2+}-EDTA complexes change in this region by several orders of magnitude (Figure 58). So, changes of pH in a system containing a protein, chelator, and Ca^{2+} will cause redistribution of Ca^{2+} between the protein and chelator due to pH dependence of the Ca^{2+} binding constant of the chelator. Figure 57C shows results of such an experiment with α-lactalbumin. The change of pH from 9 to 6 causes a blue spectral shift and a decrease of fluorescence yield of the protein. These changes are caused by saturation of α-lactalbumin by Ca^{2+} ions that were complexed with EGTA at alkaline pH values. In this case, the fitting of the experimental points for fluorescence quantum yield by the theoretical curve computed according to the Scheme 31 also allows evaluation of the Ca^{2+} binding constant of the protein. For α-lactalbumin, such evaluation gives the value $6 \times 10^8\ M^{-1}$ at 20°C. It is clearly seen that these two methods give practically the same result.

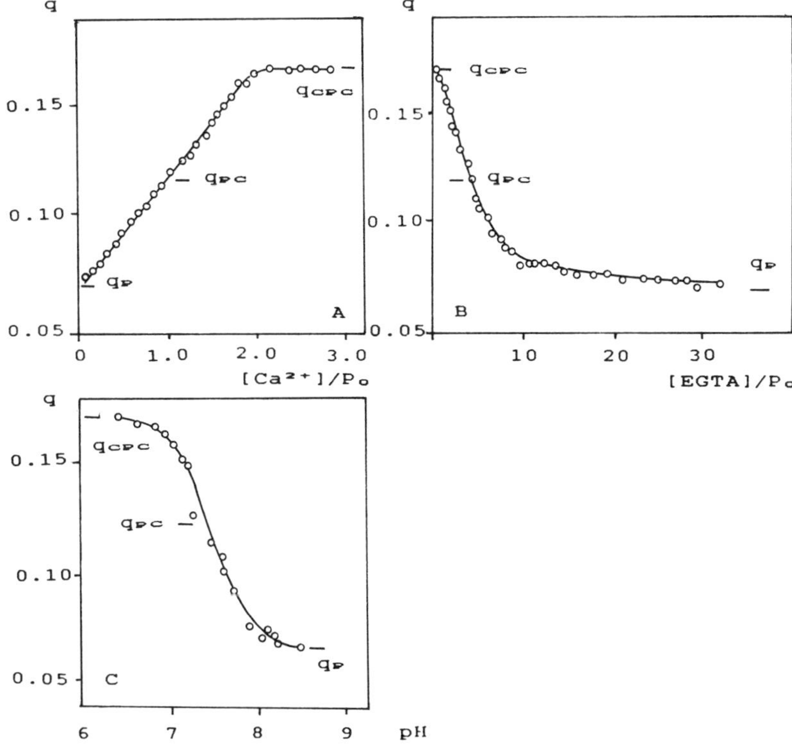

Figure 59. Pike parvalbumin, pI 5.0, 50 m*M* Tris-HCl, 20°C. (A) Fluorimetric Ca^{2+} titration; (B) fluorimetric EGTA titration of the Ca^{2+}-loaded protein, pH 8.1; (C) fluorimetric pH titration in the presence of 0.6 m*M* Ca^{2+} and 8.2 m*M* EGTA. Protein concentration P_0 = 0.25 m*M*. (From Permyakov, E. A., Medvedkin, V. N., et al., *Arch. Biochem. Biophys.*, 227, 9, 1983. With permission.)

These methods can also be used for proteins that bind more than one cation per molecule. Figures 59 and 60 show results of fluorimetric titration of pike parvalbumins, pI 5.0 (phenylalanine fluorescence) and pI 4.2 (tyrosine fluorescence), by Ca^{2+} and EGTA. The same figures show the data for pH titration of these parvalbumins in the presence of known concentrations of Ca^{2+} and EGTA.[26]

It is clearly seen from the figure that phenylalanine and tyrosine fluorescence of the parvalbumins respond to the binding

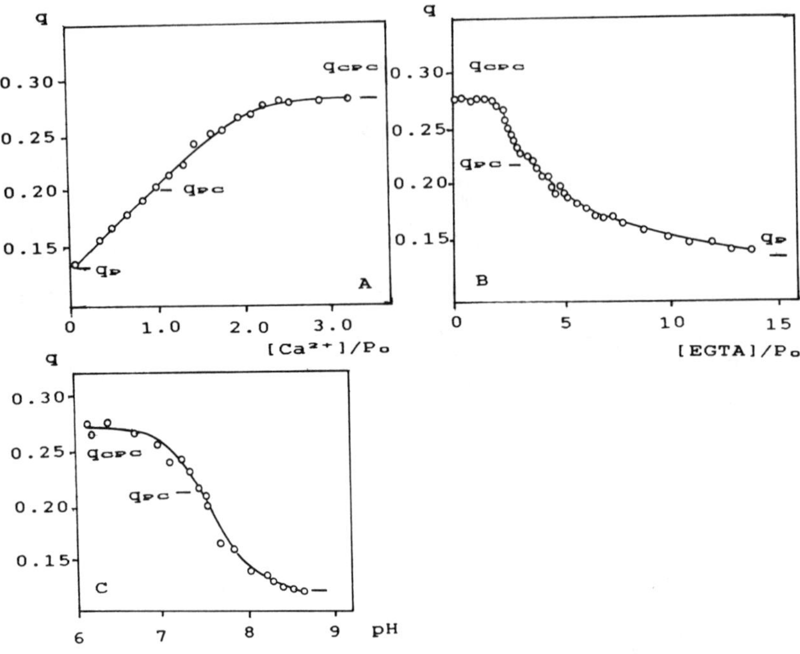

Figure 60. Pike parvalbumin, pI 4.2, 50 m*M* Tris-HCl, 20°C. (A) Fluorimetric Ca^{2+}-titration; (B) fluorimetric EGTA titration of the Ca^{2+}-loaded protein, pH 8.1; (C) fluorimetric pH titration in the presence of 0.5 m*M* Ca^{2+} and 2 m*M* EGTA. Protein concentration P_0 = 0.1 m*M*. (From Permyakov, E. A., Medvedkin, V. N., et al., *Arch. Biochem. Biophys.*, 227, 9, 1983. With permission.)

of two Ca^{2+} ions per protein molecule. Parvalbumins are assumed to bind two ions successively:

$$
\begin{aligned}
P + Ca^{2+} &\overset{K_1}{\rightleftharpoons} PCa \\
PCa + Ca^{2+} &\overset{K_2}{\rightleftharpoons} CaPCa \\
(EGTA + Ca^{2+} &\overset{K_E}{\rightleftharpoons} EGTACa)
\end{aligned}
\tag{32}
$$

Table 3. Equilibrium Ca^{2+} Binding Constants of Pike Parvalbumins Evaluated in Three Types of Experiments

	Parvalbumin (pI 5.0)		Parvalbumin (pI 4.2)	
Type of experiment	K_1 (M^{-1})	K_2 (M^{-1})	K_1 (M^{-1})	K_2(M^{-1})
Ca^{2+} titration	$>10^6$	$>10^6$	$>10^6$	2.2×10^5
EGTA titration	1.8×10^8	7.5×10^8	3.1×10^8	2.2×10^4
pH titration	6.3×10^8	4.3×10^8	2.8×10^8	6.9×10^5

From Permyakov, E. A., Kalinichenko, L. P., Medvedkin, V. N., Burstein, E. A., and Gerday, C., *Biochim. Biophys. Acta*, 749, 185, 1983. With permission.

Fluorescence quantum yields of the P, PCa, and CaPCa were determined from the curves for the Ca^{2+} titration. Values of K_1 and K_2 were determined from the fitting of theoretical curves computed according to Scheme 32 to the experimental points by variation of the constants. Table 3 contains sets of K_1 and K_2 values obtained in these three types of experiments. The relative accuracy of the binding constant evaluation from the data of the three types of the experiments is different. Constant K_1 (and K_2 in the case of parvalbumin, pI 5.0) is determined with the highest accuracy from the EGTA titration experiment, since in this case it is comparable to the Ca^{2+}-EGTA association constant. In the Ca^{2+} titration experiment, constant K_1 (and K_2 for parvalbumin, pI 5.0) is much higher than the reciprocal protein concentration. For this reason the accuracy of evaluation of this constnat in such an experiment is low. The value of K_2 for parvalbumin pI 4.2 is determined most accurately from the Ca^{2+} titration, since in this case it is comparable with reciprocal protein concentration and the curve of the fluorimetric Ca^{2+}-titration is sufficiently sloping gently and differs from the straight line with a break.

Tryptophan fluorescence allows one to obtain richer information compared with phenylalanine and tyrosine fluorescence. This is seen well in studies of Ca^{2+} binding to whiting parvalbumin which possesses a single tryptophan residue.[11]

Results of Ca^{2+} titration of Ca^{2+}-free whiting parvalbumin are shown in Figure 61. The parameters λ, $\Delta\lambda$, and q are plotted against the logarithm of relative Ca^{2+} concentration. Gradual

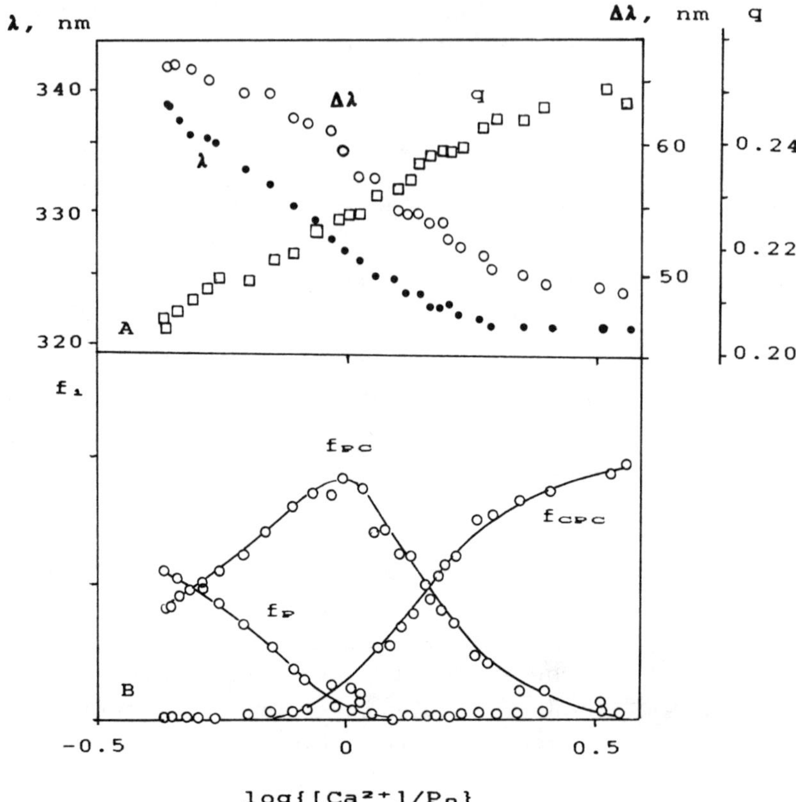

Figure 61. Spectrofluorimetric Ca^{2+}-titration of whiting parvalbumin. Protein concentration of P_0 = 20 μ*M*; 50 m*M* Hepes, pH 7.5; 20°C. (A) Spectrum position, λ; and width Δλ; q — fluorescence yield. (B) Populations of P, PCa, and CaPCa states. Points are experimental; curves are theoretical. (From Permyakov, E. A., Yarmolenko, V. V., et al., *Eur. J. Biochem.*, 109, 307, 1980. With permission.)

increase of Ca^{2+} concentration causes a narrowing of the fluorescence spectrum, its blue shift, and an increase of fluorescence yield. The curves reach plateau at $[Ca^{2+}]/P_0 \approx 2$. This shows that the binding of Ca^{2+} causes a conformational change wherein during the course of which the tryptophan residue is transferred from the protein surface to a rigid hydrophobic environment.

Fluorescent phase plots for these data (Figure 62) demonstrate a break, suggesting the existence of a stable intermediate state

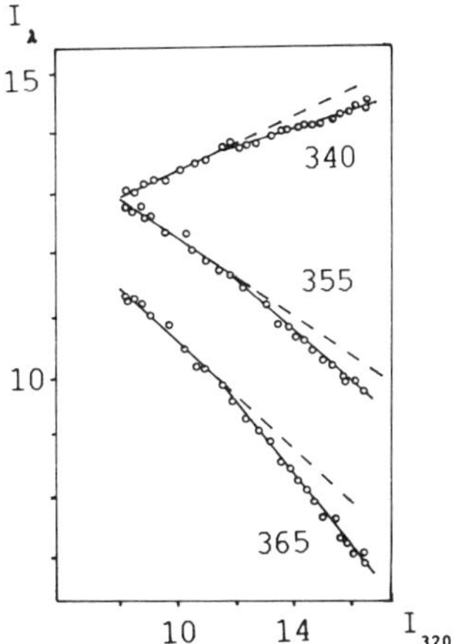

Figure 62. Fluorescent phase plots corresponding to the Ca^{2+} titration of whiting parvalbumin (Figure 61). (From Permyakov, E. A., Yarmolenko, V. V., et al., *Eur. J. Biochem.*, 109, 307, 1980. With permission.)

which seems to correspond to the protein with one bound Ca^{2+}. Fluorescence spectra of P, PCa, and CaPCa states are shown in Figure 63.

Fluorescence spectra of the protein at intermediate concentrations of Ca^{2+} were approximated by a sum of emission spectra of P, PCa, and CaPCa with contributions a_p, a_{pCa}, and a_{CapCa}. The population of the ith state is f_i:

$$f_P = [P]/P_0, \quad f_{PCa} = [PCa]/P_0, \quad f_{CaPCa} = [CaPCa]/P_0 \qquad (33)$$

where P_0 is total protein concentration.

In the case in which a spectrum consists of two components, for example, components corresponding to the emission of P and PCa,

$$a_{PCa} = f_{PCa} \cdot q_{PCa} / q \quad \text{and} \quad a_P = f_P \cdot q_P / q \qquad (34)$$

where q_{pCa} and q_p are fluorescence yields of PCa and P, and q is mean fluorescence yield of the protein sample at a given Ca^{2+} concentration.

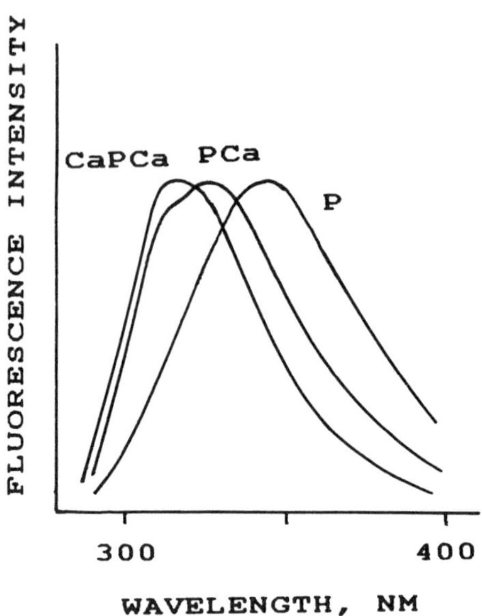

Figure 63. Fluorescence spectra of the P, PCa, and CaPCa states of whiting parvalbumin. (From Permyakov, E. A., Yarmolenko, V. V., et al., *Eur. J. Biochem.*, 109, 307, 1980. With permission.)

$$a_{PCa} + a_P = 1 \quad \text{and} \quad f_{PCa} + f_P = 1 \tag{35}$$

or

$$1/q = (1/q_{PCa} - 1/q_P) \cdot a_{PCa} + 1/q_P \tag{36}$$

q_{pCa} and q_p can be determined from the plot of $1/q$ vs. a_{pCa} (Figure 64). A similar procedure can be done for the other part of the titration curve where the spectra consist of the PCa and CaPCa components. The knowledge of q_p, q_{pCa}, and q_{CapCa} allows determination of f_p, f_{pCa}, and f_{CapCa}. Figure 61B shows the dependence of f_p, f_{pCa}, and f_{CapCa} on the logarithm of relative Ca^{2+} concentration. The points in the plot are experimental, while the curves are theoretical ones which were computed according to the scheme of successive binding of two Ca^{2+} ions to the parvalbumin molecule and fit to the experimental points by

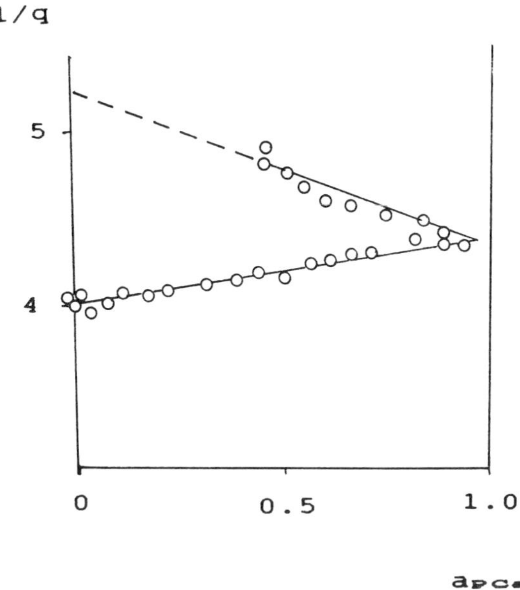

Figure 64. Evaluation of fluorescence yields of the P, PCa, and CaPCa states of whiting parvalbumin according to Equation 36. (From Permyakov, E. A., Yarmolenko, V. V., et al., *Eur. J. Biochem.*, 109, 307, 1980. With permission.)

variation of K_1 and K_2. The fit gives $K_1 > 10^6$ M^{-1} and $K_2 = 6 \times 10^6$ M^{-1}.

In order to evaluate K_1, one should carry out the experiment on EGTA titration of the Ca^{2+}-loaded protein. Figure 65 shows the results of such an experiment. The fitting of the population data of the P, PCa, and CaPCa states by theoretical curves computed according to the scheme of competition of the protein and EGTA for Ca^{2+} gives values $K_1 = 7 \times 10^8$ M^{-1} and $K_2 = 6 \times 10^6$ M^{-1}.

An analysis of the pH dependence of fluorescence parameters of the parvalbumin measured in the presence of known Ca^{2+} and EGTA concentrations can serve as an additional method of determination of the binding constants. Results of such an experiment are shown in Figure 66. This method gives the values $K_1 = 2 \times 10^8$ M^{-1} and $K_2 = 2 \times 10^6$ M^{-1}.

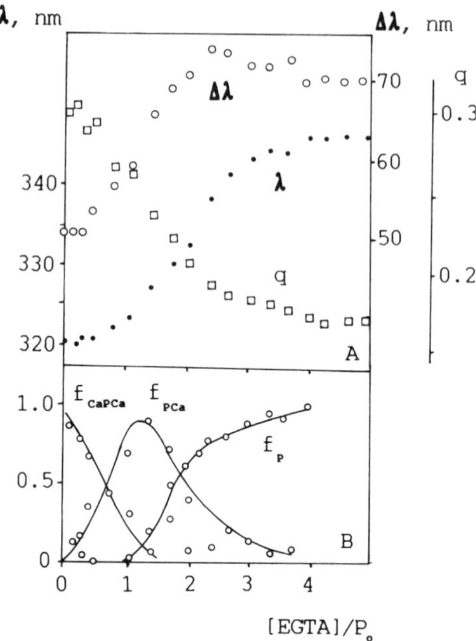

Figure 65. Spectrofluorimetric EGTA titration of Ca^{2+}-loaded whiting parvalbumin (50 m*M* Hepes, pH 8.8, 20°C, protein concentration P_0 = 20 μM. (A) Fluorescence spectrum position, λ; width, Δλ; and quantum yield, q. (B) Populations of the P, PCa, and CaPCa states. (From Permyakov, E. A., Yarmolenko, V. V., et al., *Eur. J. Biochem.*, 109, 307, 1980. With permission.)

So, whiting parvalbumin binds two Ca^{2+} ions with constants $K_1 = 5 \times 10^8\ M^{-1}$ and $K_2 = 6 \times 10^6\ M^{-1}$.

Let's consider some examples of evaluation of binding constants for proteins when $P_0 << 1/K$. Figure 67 shows results of the fluorimetric study of the binding of ATP and ADP to bovine α-lactalbumin at 37°C.[31] Gradual addition of the nucleotides changes both tryptophan fluorescence yield and spectrum position. The spectrum shifts to shorter wavelengths in the case of the Ca^{2+}-loaded protein and to longer wavelengths in the case of ion-free α-lactalbumin. The curves reach a plateau at high concentrations of the nucleotides. The main spectral changes occur at millimolar concentrations of the nucleotides, while the protein concentration is in the micromolar range. This means that the affinity of

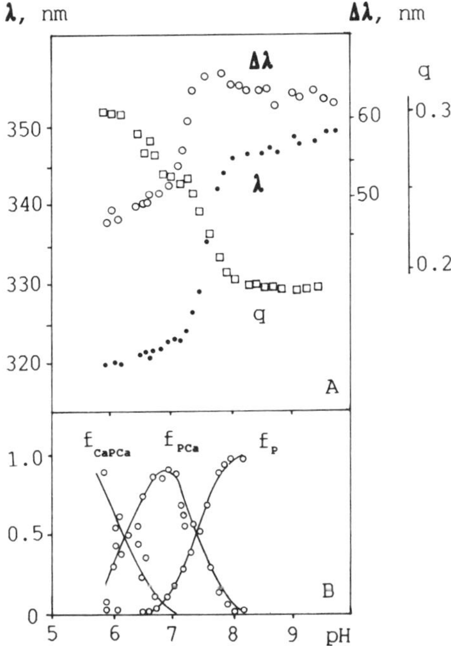

Figure 66. Spectrofluorimetric pH titration of whiting parvalbumin (77 μ*M*) in the presence of 170 and 625 μ*M* EGTA (50 m*M* Hepes, 20°C) (A) Fluorescence spectrum position, λ; width, Δλ; and quantum yield. (B) Populations of the P, PCa, and CaPCa states. (From Permyakov, E. A., Yarmolenko, V. V., et al., *Eur. J. Biochem.*, 109, 307, 1980. With permission.)

α-lactalbumin for the nucleotides is rather low. We cannot determine a binding stoichiometry in this case. An evaluation of the binding constant on the basis of the simple one-site model provides 1050 and 130 M^{-1} values for the binding of ATP to Ca^{2+}-loaded and Ca^{2+}-free protein, respectively. Similar values for ADP are 130 and 180 M^{-1}, respectively.

Figure 68 shows the results of spectrofluorimetric titration of α-lactalbumin by UDP-galactose, the substrate of the lactose synthase reaction.[31] It is clearly seen that the spectral changes in this case occur also at millimolar concentrations, while protein concentration is 20 μ*M*. It is worthwhile to note the sigmoidal dependence of fluorescence quantum yield on UDP-galactose

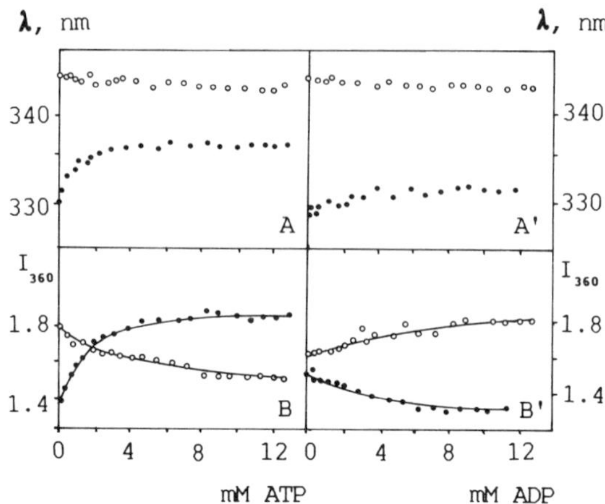

Figure 67. Spectrofluorimetric ATP (A,B) and ADP (A′,B′) titration of Ca^{2+}-loaded (•) and Ca^{2+}-free (○) α-lactalbumin. Protein concentration 0.02 m*M*. (A,A′) λ fluorescence spectrum position; (B,B′) fluorescence intensity at 360 nm. (From Permyakov, E. A. and Kreimer, D. I., *Gen. Physiol. Biophys.*, 5, 377, 1986. With permission.)

concentration for the Ca^{2+}-loaded protein. This means that either the binding of the first UDP-Gal molecules gives no fluorescent response, or the binding is actually cooperative. In any case, such a curve suggests that the protein binds more than one UDP-Gal molecule. An apparent UDP-Gal binding constant can be evaluated by approximating the experimental data by the following scheme:

$$P + mN \overset{K}{\rightleftharpoons} PN_m \tag{37}$$

where P is the protein and N is UDP-Gal. This scheme suggests a simultaneous binding of m UDP-Gal molecules to the protein, i.e., the protein can be in two states: P and PN_m. It is evident that this scheme is only a rough approximation to the real one. Fitting of a theoretical curve computed according to Scheme 37 to the experimental points gives values $K = 10^3\ M^{-1}$ and m = 3.8.

In the case of the metal-free protein, we can use

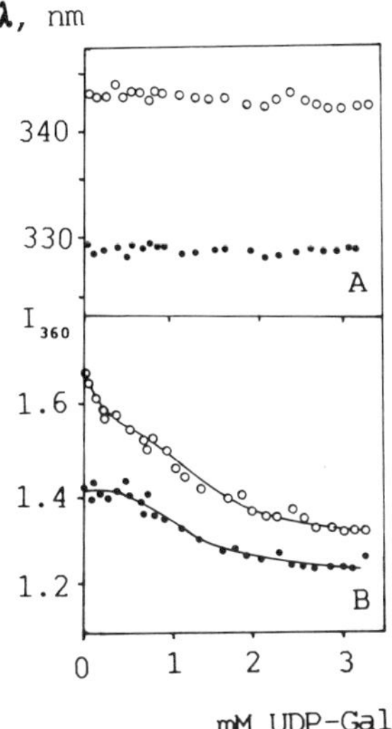

Figure 68. Spectrofluorimetric UDP-Gal titration of Ca^{2+}-loaded (•) and Ca^{2+}-free (○) α-lactalbumin. Protein concentration 0.03 m*M*. (A) Spectrum position; (B) fluorescence intensity at 360 nm. (From Permyakov, E. A. and Kreimer, D. I., *Gen. Physiol. Biophys.*, 5, 377, 1986. With permission.)

$$\begin{aligned} P + N &\overset{K_1}{\rightleftharpoons} PN \\ PN + mN &\overset{K_2}{\rightleftharpoons} PN_{m+1} \end{aligned} \tag{38}$$

which differs from Scheme 37 by an additional equation describing the initial part of the curve. The fitting gives the following values: $K_1 = 8.6 \times 10^3\ M^{-1}$, $K_2 = 770\ M^{-1}$, m = 3.2.

Sometimes it is convenient to use fluorescent phase plots for determination of binding stoichiometry. Figure 69 shows results of spectrofluorimetric Na^+ and K^+ titration of α-lactalbumin at 20°C. The spectral changes occur at decimolar concentrations of the ions, while protein concentration is 15 μ*M*, i.e., in this case, $P_0 \ll 1/K$. The K^+ and Na^+ ions cause quantitatively different fluorescent changes, which suggests that these changes are caused

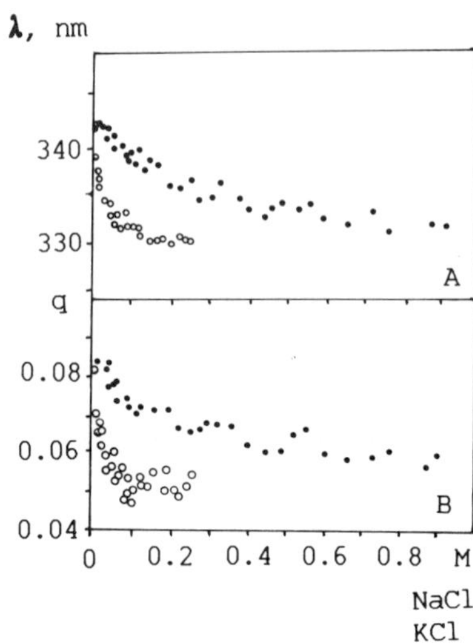

Figure 69. Spectrofluorimetric Na^+ and K^+ titration of α-lactalbumin. (A) Fluorescence spectrum position; (B) fluorescence quantum yield. Protein concentration 0.015 m*M*. (○) NaCl; (•) KCl. (From Permyakov, E. A., Morozova, L. A., and Burstein, E. A., *Biophys. Chem.*, 21, 21, 1985. With permission.)

not by trivial effects of ionic strength, but by cation binding to the protein. Figure 70 shows fluorescent phase plots corresponding to the data presented in the preceding figure. It is clearly seen that the plot for KCl is a straight line, while the plot for NaCl has a break. This shows that the protein binds at least two Na^+ ions per molecule.

XIII. METHOD OF SELECTIVE FLUORESCENCE QUENCHING

The method of selective fluorescence quenching is used for studies of location of chromophores in protein molecules.[3,15] The method is based on the fact that quencher molecules which deactivate excited states only during collisions will quench the

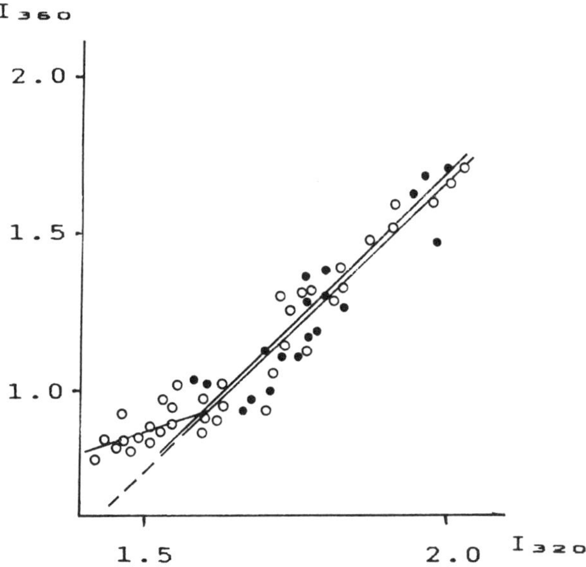

Figure 70. Fluorescent phase plots corresponding to the Na^+ and K^+ titration of α-lactalbumin. (○) NaCl; (•) KCl. (From Permyakov, E. A., Morozova, L. A., Burstein, E. A., *Biophys. Chem.*, 21, 21, 1985. With permission.)

fluorescence of external, solvent-accessible chromophores more effectively than they will the emission of buried ones. Quencher molecules hardly penetrate to the internal regions of proteins. In this respect the method of selective fluorescence quenching is similar to the method of selective chemical modification or the method of perturbants.

As was mentioned above, fluorescence quenching of the second kind, i.e., as a result of collisions of quencher molecules with excited chromophores, obeys the Stern-Volmer equation. If a solution contains molecules of n forms of fluorescent substances with different quenching constants K_i and different contributions to the total fluorescence spectrum a_i, ($\Sigma\, a_i = 1$), the Stern-Volmer equation will look like this:

$$I = \frac{a_1 \cdot I_0}{1 + K_1 \cdot C} + \frac{a_2 \cdot I_0}{1 + K_2 \cdot C} + \cdots + \frac{a_n \cdot I_0}{1 + K_n \cdot C} \tag{39}$$

where I_0 is fluorescence intensity in the absence of quencher and C is quencher concentration. For two fluorescent forms,

$$I = \frac{a_1 \cdot I_0}{1 + K_1 \cdot C} + \frac{a_2 \cdot I_0}{1 + K_2 \cdot C} \qquad (40)$$

One can use a value called the quenching efficiency, Q:

$$Q = \frac{I_0}{I} - 1 = \frac{(a_1K_1 + a_2K_2)C + K_1K_2C^2}{1 + (a_1K_2 + a_2K_1)C} \qquad (41)$$

There may be some deviations from the Stern-Volmer equation. For example, quencher molecules may deactivate excited chromophores by forming nonfluorescent complexes with them in excited states. Such quenching is called *static quenching*. Static quenching distorts the Stern-Volmer plots. In contrast to dynamic quenching, static quenching does not decrease fluorescence lifetime. In order to take into account static quenching, one can use a modified Stern-Volmer equation:

$$I = \frac{I_0 \exp(-\omega C)}{1 + KC} \qquad (42)$$

where ω characterizes an "action sphere" of the quencher.

Another factor which causes deviations from the Stern-Volmer equation is the possible binding of a quencher to proteins. This can increase both static and dynamic quenching. Moreover, the binding of the quencher can change protein conformation which, in turn, can cause changes in location of protein chromophores and their accessibility to solvent. The absence of binding of a quencher to a protein is one of the main conditions of applicability of the method of selective fluorescence quenching.

Since the method of selective fluorescence quenching requires the use of rather high concentrations of quenchers which are often ionic, one should be sure that the ionic strength created by a quencher does not change protein structure. If such effects take place, all experiments on fluorescence quenching must be carried

Table 4. Stern-Volmer Constants K and Bimolecular Rate Constants k_q for Fluorescence Quenching of Aromatic Amino Acids in Aqueous Solution by Some Quenchers[2,3]

Quencher	$K(M^{-1})$				k_q ($\times 10^9$ $M^{-1}c^{-1}$)			
Amino acid	Cs^+	I^-	SCN^-	NO_3^-	Cs^+	I^-	SCN^-	NO_3^-
Phenylalanine	1.35	—	20.0	—	0.20	—	3.84	
Tyrosine	4.6	19.9	3.0	—	1.43	0.42	0.97	
Tryptophan	2.8	9.1	0	30.0	1.0	3.25	0	10.7

out at a constant ionic strength using the ionic quencher with some nonquenching ion.

Most commonly used ionic quenchers of fluorescence of tryptophan, tyrosine, and phenylalanine residues are Cs^+, I^-, SCN^-, and NO_3^-. Acrylamide is usually used as a nonionic quencher. Stern-Volmer constants and bimolecular rate constants, k_q, for fluorescence quenching of free amino acids in aqueous solutions by ionic quenchers are collected in Table 4.

It should be noted that some quenchers, for example, NO_3^-, absorb light both at the excitation wavelength and in the region of protein fluorescence. In such cases one should correct the fluorescence spectra for the screening and reabsorption effects.

The parameters of fluorescence quenching can be determined with sufficient accuracy only for proteins containing only two types of chromophores. The experiment requires measurement of fluorescence quantum yield or fluorescence intensity as a function of quencher concentration. Then these data are represented in the form $Q = I_0/I - 1$ vs. quencher concentration, C. In the case of one type of emitting sites, the slope of the obtained straight line gives the Stern-Volmer constant (Figure 71). The data can be represented also as $I_0/(I_0 - 1)$ vs. $1/C$:

$$\frac{I_0}{I_0 - I} = \frac{1}{a \cdot K \cdot C} + \frac{1}{a} \tag{43}$$

where a is the fraction of chromophores accessible to quencher. For proteins containing two or more sites of quenching, such

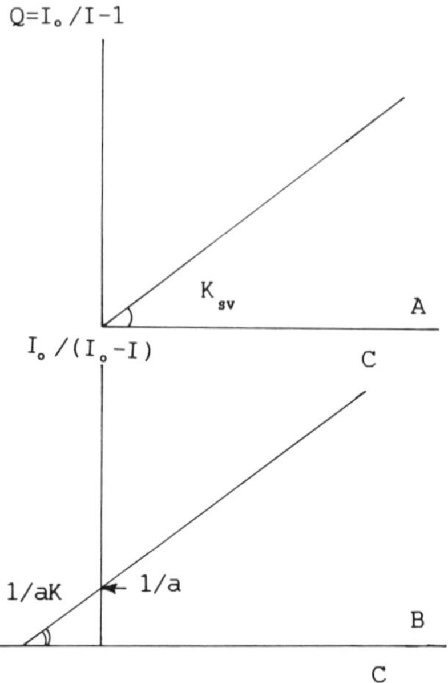

Figure 71. Two different ways of representation of data for fluorescence quenching.

plots will deviate from straight lines. Nevertheless, by using methods of computer fitting of theoretical curves computed according to a chosen quenching model to experimental points by variation of the quenching parameters, one can evaluate these parameters rather accurately. Figure 72 shows data on fluorescence quenching of the laminarinase-laminarin complex by I^- ions. The curve in the figure is theoretical and was computed according to Equation 40. The best fit is achieved when $K_1 = 0.48 \pm 0.14\ M^{-1}$, $K_2 = 40.5 \pm 7.2\ M^{-1}$, and $a_1 = 0.52 \pm 0.03$.

It is usually hard to compare Stern-Volmer constants determined for different chromophores in proteins. For this reason a parameter L is used as a measure of accessibility of a chromophore to a quencher:[4]

$$L = \frac{K_P}{K_{AA}} \cdot \frac{q^0_{AA}}{q^0_P} \qquad (44)$$

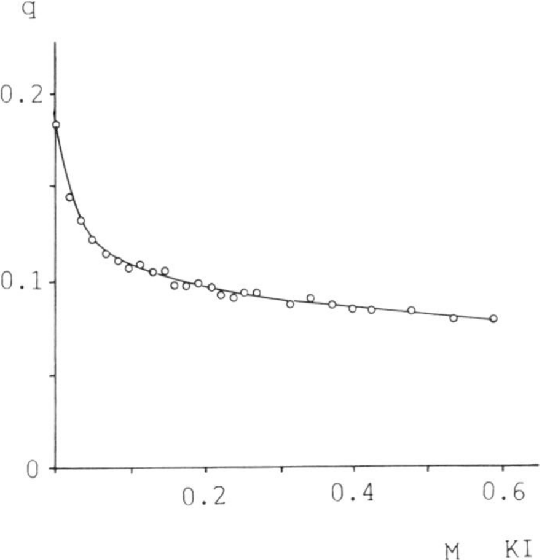

Figure 72. Fluorescence quenching of laminarinase-laminarin complex by I^-. Points are experimental, and the curve is a theoretical one computed according to Equation 40 and fitted to experimental points.

where K_p and K_{AA} are Stern-Volmer constants for the protein and free amino acid in solution; q_p^0 and q_{AA}^0 are fluorescence yields of the protein and free amino acid in the absence of the quencher, respectively. If $K = \tau_f \cdot k_q$ and $q_0 = \tau_f \cdot \tau_f$ (τ_f is fluorescence lifetime, k_f is the rate constant of the fluorescent transition, and k_q is the bimolecular rate constant of the interaction of the quencher with the protein), then

$$L = \frac{k_q^P}{k_q^{AA}} \tag{45}$$

Since k_q is limited by diffusion, L is a relative coefficient of diffusion of the quencher to the chromophore located in the protein structure. For surface chromophores, L lies within the region from 0.4 to 0.6, while for buried chromophores, L <0.10.

In some cases, L can be more than 1, which is explained by effects of electrostatic attraction of the quencher by oppositely charged groups in the environment of the chromophore. The use of a pair of quenchers, anionic and cationic, allows qualitative evaluation of the electrostatic situation in the chromophore environment.

The successful use of the method of selective quenching greatly depends on the choice of quenchers. The main requirements for quenchers are

1. Fluorescence quenching upon collision with excited chromophores.
2. High efficiency of quenching.
3. Absence of any specific interactions with proteins.
4. Absence of nonspecific "solution" of the quencher in the hydrophobic part of the protein globule.

XIV. **LOW-TEMPERATURE LUMINESCENCE OF PROTEINS**

As was mentioned above, fluorescence spectra of proteins at low temperatures, for example, at the temperature of liquid nitrogen (–196°C), are more complex compared with those measured at temperatures above 0°C. At low temperatures tryptophan fluorescence spectra of many proteins display a vibrational structure, and the spectra at –196°C have a shorter wavelength position compared with those at temperatures above 0°C. Moreover, all proteins at low temperatures possess phosphorescence, long-lived emission, in the region from 400 to 500 nm.

Figure 73 shows luminescence spectra of various proteins in frozen aqueous solution at –196°C. Among these proteins are azurin, asparaginase, and β-lactoglobulin. A vibrational structure in the phosphorescence spectra is clearly seen. Positions of the maxima in the phosphorescence spectra can vary within 10 nm. Changes in location of tryptophan residues in protein molecules result in changes of position of maxima of both fluorescence and phosphorescence, but any clear correlation between the location of tryptophan residues in a protein and position of phosphorescence peaks was not found. For example, the denaturation of chymotrypsinogen by urea shifts its phosphorescence

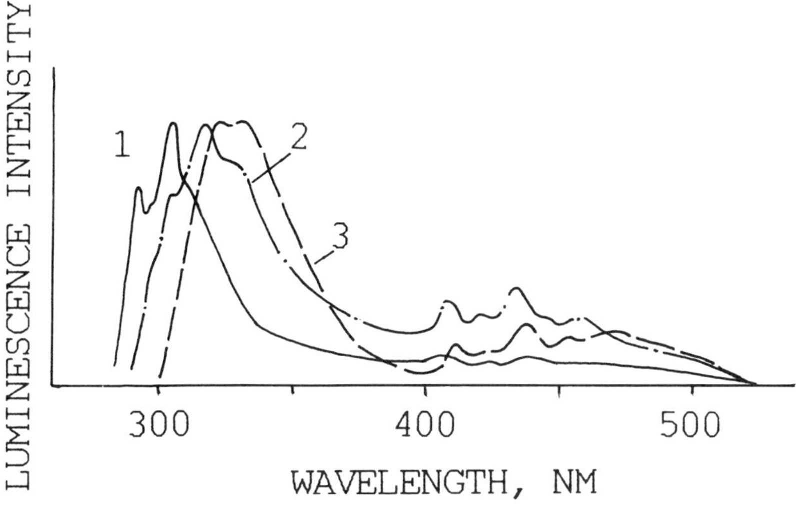

Figure 73. Emission spectra of frozen aqueous solutions of azurin (1), asparaginase (2), and β-lactoglobulin (3) at –196°C.

peaks to shorter wavelengths (Figure 74A), while the urea denaturation of papain practically does not affect the position and shape of its phosphorescence spectrum (Figure 74B). Nevertheless, urea induces a small red shift of the fluorescence spectrum for both proteins.

Phosphorescence lifetimes of most native proteins lie in the region from 4 to 7 s. Protein denaturation has practically no effect on phosphorescence lifetimes. Phosphorescence lifetimes of proteins with a quenched phosphorescence (low ratio of phosphorescence to fluorescence yields) are slightly shorter, as for example, for azurin (3.4 s), lysozyme (3.8 s), or α-lactalbumin (3.8 s). For lysozyme and α-lactalbumin, this could be explained by a quenching effect of neighboring disulfide groups.

The ratio of phosphorescence to fluorescence for most native proteins in the absence of quenchers is within the range of 0.3 to 1.0.

Decay curves for tryptophan phosphorescence of some proteins contain a short-lived component (τ <0.5 s) which contributes up to 10 to 50% of the total phosphorescence. The proteins in the presence of external quenchers such as Cs^+ and I^- display an

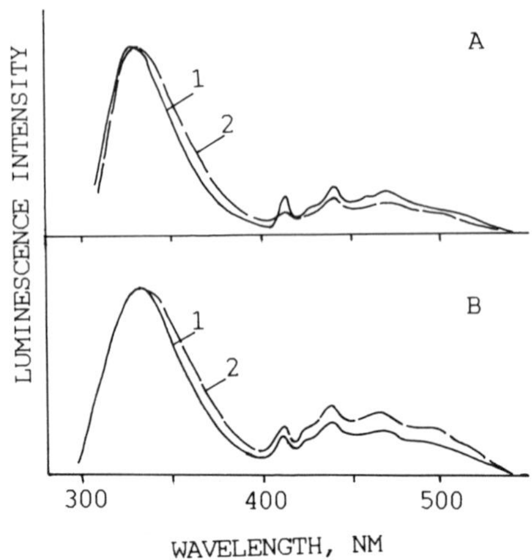

Figure 74. Normalized luminescence spectra of native (1) and urea-denatured (2) chymotrypsin (A) and papain (B) at –196°C.

especially large contribution from this component. The ions of the quenchers in frozen systems seem to be located in contact with protein chromophores, and since they are heavy atoms, they increase spin-orbit coupling and facilitate forbidden transitions: intersystem crossing $S_1 \rightsquigarrow T_1$ and radiative (phosphorescence) and nonradiative $T_1 \rightsquigarrow S_0$ transitions. As a result, phosphorescence yield increases, while phosphorescence lifetime decreases.

Figure 75 shows normalized luminescence spectra of chymotrypsinogen in the absence and in the presence of Cs^+. In the presence of the quencher, the phosphorescence to fluorescence ratio becomes more than 1. It should be noted that at low temperatures quenchers affect luminescence of both surface and buried tryptophans equally well. This seems to be explained by the fact that upon the freezing of the protein solution, the quencher is excluded from the ice crystals and concentrated in very high concentrations near protein molecules, and that a fraction of the quencher molecules seem to penetrate inside the protein structure.

Spectra of tyrosine fluorescence of proteins at –196°C have a shorter wavelength position in comparison with those measured

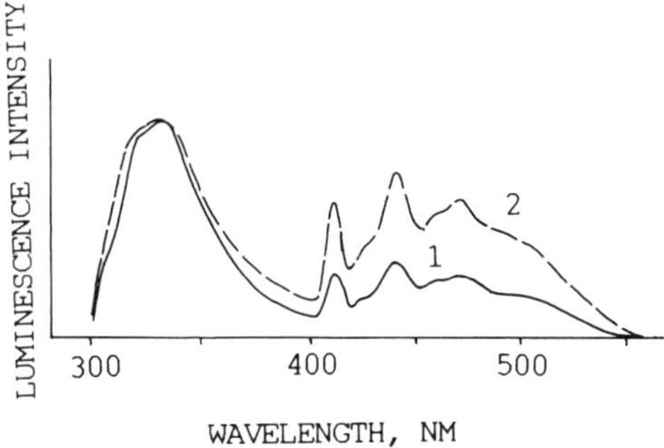

Figure 75. Normalized luminescence spectra of chymotrypsinogen in the absence (1) and in the presence of 0.5 *M* CsCl (2) at –196°C.

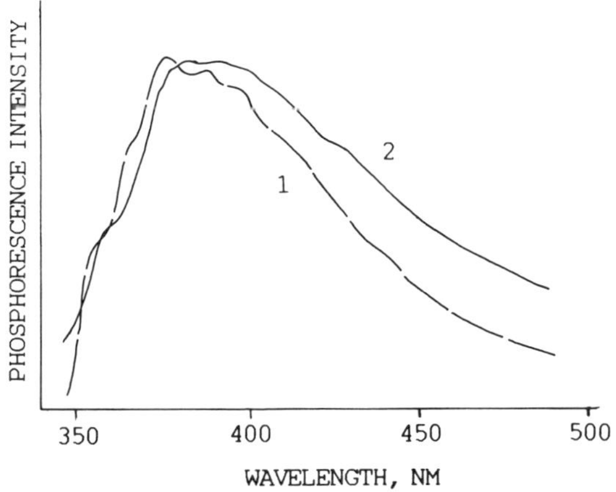

Figure 76. Phosphorescence spectra of tyrosine in 1 *M* NaCl (1) and carp parvalbumin, pI 3.95 (2), at –196°C.

at temperatures above 0°C. Phosphorescence spectra of tyrosine residues located in contact with disulfide groups display maxima in the region from 410 to 420 nm, while tyrosine residues located in other environments have shorter wavelength spectra similar to the phosphorescence spectrum of tyrosine in frozen water-salt solutions (maximum at 387 nm). Figure 76 shows the phosphorescence spectrum of carp parvalbumin, pI 3.95, which contains one tyrosine and nine phenylalanine residues. The lifetime of

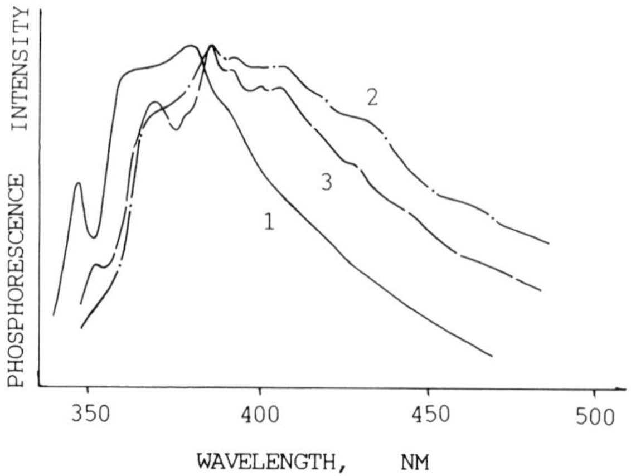

Figure 77. Phosphorescence spectra of phenylalanine in 1 *M* NaCl (1) and hake parvalbumin (2) and carp parvalbumin, pI 4.47 (3), at –196°C.

tyrosine phosphorescence of the parvalbumin is 2.5 s, which is close to the phosphorescence lifetime of free tyrosine in frozen water-salt solution (2.4 s).

Spectra of phenylalanine fluorescence at –196°C possess vibrational peaks, the positions of which coincide with their positions at room temperature. The phosphorescence spectra of phenylalanine residues in proteins are similar in shape to the spectrum of free phenylalanine in water-salt solution. It has a rich vibrational structure, but is slightly shifted towards longer wavelengths. Figure 77 shows spectra of phenylalanine phosphorescence of carp and hake parvalbumins. These proteins do not contain either tyrosine or tryptophan residues, but possess ten phenylalanine residues per molecule.

REFERENCES

1. Teale, F. W. J. and Weber, G., *Biochem. J.*, 65, 467, 1957.
2. Burstein, E. A., Permyakov, E. A., Yashin, V. A., Burkhanov, S. A., and Finazzi-Agro, A., *Biochem. Biophys. Acta*, 491, 155, 1977.
3. Burstein, E. A., *Luminescence of Protein Chromophores*, Vol. 6, Model Studies. Science and Technology Results. Biophysics, VINITI, Moscow, 1976.

4. Burstein, E. A., *Intrinsic Protein Luminescence,* Vol. 7, Origin and Applications. Science and Technology Results. Biophysics, VINITI, Moscow, 1977.
5. Burstein, E. A., Vedenkina, N. S., and Ivkova, M. N., *Photochem. Photobiol.,* 18, 263, 1973.
6. Permyakov, E. A. and Shnyrov, V. L., *Biophys. Chem.,* 18, 145, 1983.
7. Grishchenko, V. M., Emelyanenko, V. I., Ivkova, M. N., Bezborodova, S. I., and Burstein, E. A., *Bioorg. Chem.,* 2, 207, 1976.
8. Permyakov, E. A., Cooperative freezing of internal mobility of proteins, in *Equilibrium Dynamics of Native Protein Structure,* Pushchino, 1977, 83.
9. Permyakov, E. A., Yarmolenko, V. V., Kalinichenko, L. P., Morozova, L. A., and Burstein, E. A., *Biofizika,* 27, 380, 1982.
10. Permyakov, E. A., Yarmolenko, V. V., Kalinichenko, L. P., Morozova, L. A., and Burstein, E. A., *Biochem. Biophys. Res. Commun.,* 100, 191, 1981.
11. Permyakov, E. A., Yarmolenko, V. V., Emelyanenko, V. I., Burstein, E. A., Gerday, C., and Closset, J., *Eur. J. Biochem.,* 109, 307, 1980.
12. Emelyanenko, V. I. and Burstein, E. A., in *Proc. 5th Int. Conf. Spectroscopy of Biopolymers,* Kharkov, 1984, 83.
13. Permyakov, E. A., Burstein, E. A., Emelyanenko, V. I., Alexandrov, Y. M., Glagolev, K. V., Makhov, V. N., Syreishchikova, T. N., and Yakimenko, M. N., *Biofizika,* 28, 393, 1983.
14. Cundall, R. B. and Dale, R. E., *Time Resolved Fluorescence Spectroscopy in Biochemistry and Biology,* Plenum Press, New York, 1983.
15. Lakowicz, J. R., *Principles of Fluorescence Spectroscopy,* Plenum Press, New York, 1990.
16. Permyakov, E. A., Ostrovsky, A. V., Burstein, E. A., Pleshanov, P. G., and Gerday, C., *Arch. Biochem. Biophys.,* 240, 781, 1985.
17. Szabo, A. G., Stepanik, T. M., Wayner, D. M., and Young, N. M., *Biophys. J.,* 41, 233, 1983.
18. Bushueva, T. L., Busel, E. P., and Burstein, E. A., *Biochem. Biophys. Acta,* 534, 141, 1978.
19. Bushueva, T. L., Busel, E. P., and Burstein, E. A., *Arch. Biochem. Biophys.,* 204, 161, 1980.
20. Permyakov, E. A. and Burstein, E. A., *Biophys. Chem.,* 19, 265, 1984.
21. Kaye, G. W. and Laby, T. H., *Tables of Physical and Chemical Constants,* Longmans, Green, London, 1959.
22. Permyakov, E. A., Medvedkin, V. N., Kalinichenko, L. P., and Burstein, E. A., *Arch. Biochem. Biophys.,* 227, 9, 1983.
23. Ivkova, M. N., Vedenkina, N. S., and Burstein, E. A., *Mol. Biol. (U.S.S.R.),* 5, 214, 1971.
24. Permyakov, E. A., Morozova, L. A., and Burstein, E. A., *Biophys. Chem.,* 21, 21, 1985.

25. Permyakov, E. A., Shnyrov, V. L., Kalinichenko, L. P., and Orlov, N. Y., *Biochem. Biophys. Acta,* 830, 288, 1985.
26. Permyakov, E. A., Kalinichenko, L. P., Medvedkin, V. N., Burstein, E. A., and Gerday, C., *Biochem. Biophys. Acta,* 749, 185, 1983.
27. Permyakov, E. A., Yarmolenko, V. V., Burstein, E. A., and Gerday, C., *Biophys. Chem.,* 15, 19, 1982.
28. Scatchard, G., *Ann. N.Y. Acad. Sci.,* 51, 660, 1949.
29. Cantor, R. and Schimmel, P. R., *Biophysical Chemistry,* W. H. Freeman, San Francisco, 1979.
30. Schwarzenbach, G. and Flaschka, H., *Die Komplexonometrische Titration,* Ferdinand Enke Verlag, Stuttgart, 1965.
31. Permyakov, E. A. and Kreimer, D. I., *Gen. Physiol. Biophys.,* 5, 377, 1986.

Index

G

H

I

K

N

O

P

Q

R

S

T

U

V

W

Z